Haaff's Practical Dehorner
or, Every Man His Own Dehorner

by H.H. Haaff

with an introduction by Jackson Chambers

This work contains material that was originally published in 1888.

This publication is within the Public Domain.

This edition is reprinted for educational purposes
and in accordance with all applicable Federal Laws.

Introduction Copyright 2018 by Jackson Chambers

Self Reliance Books

Get more historic titles on animal and stock breeding, gardening and old fashioned skills by visiting us at:

http://selfreliancebooks.blogspot.com/

Introduction

I am pleased to present another title in the "Cattle" series.

The work is in the Public Domain and is re-printed here in accordance with Federal Laws.

As with all reprinted books of this age that are intended to perfectly reproduce the original edition, considerable pains and effort had to be undertaken to correct fading and sometimes outright damage to existing proofs of this title. At times, this task is quite monumental, requiring an almost total "rebuilding" of some pages from digital proofs of multiple copies. Despite this, imperfections still sometimes exist in the final proof and may detract from the visual appearance of the text.

I hope you enjoy reading this book as much as I enjoyed making it available to readers again.

Jackson Chambers

PREFACE.

The universal feeling of good will with which "Haaff on Dehorning Cattle" has been received among all the farmers and cattle-men everywhere, and the unalloyed satisfaction which the operation of "dehorning cattle" has given in every one of the thousands upon thousands of farms and ranches where the author's directions and tools have been used and followed, together with the resulting benefits bestowed upon both man and the brute creation, the many human lives saved and the thousands of pockets benefited, and, more than all, the hundreds of letters from one end of the continent to the other, and from Europe, South America and Australia, demanding to be given the reason why, and the time when, and the place where, and the way how "dehorning cattle" of all ages, sizes, conditions and kinds may be best done, these and the persistent determination of the author to make "Every Man his own Dehorner," practical and perfect—an adept in the art—these constitute the apology for offering "Haaff's Practical Dehorner, or Dehorning Cattle Illustrated," with full cuts and ample description of every possible phase of the operation and all the attendant circumstances, to the public.

THE AUTHOR.

AN APOLOGY

Is perhaps due the readers of this volume for errors in the composition or in the typography of the work. The author will accept all criticisms, if any there are on this line personally, for in the attempt to issue this book during the month of March, it has so crowded us all that little chance has been given to review these pages. Please excuse any manifest errors and charge them to account of our crowded time and a desire to serve the public at as early a day as possible.

THE AUTHOR.

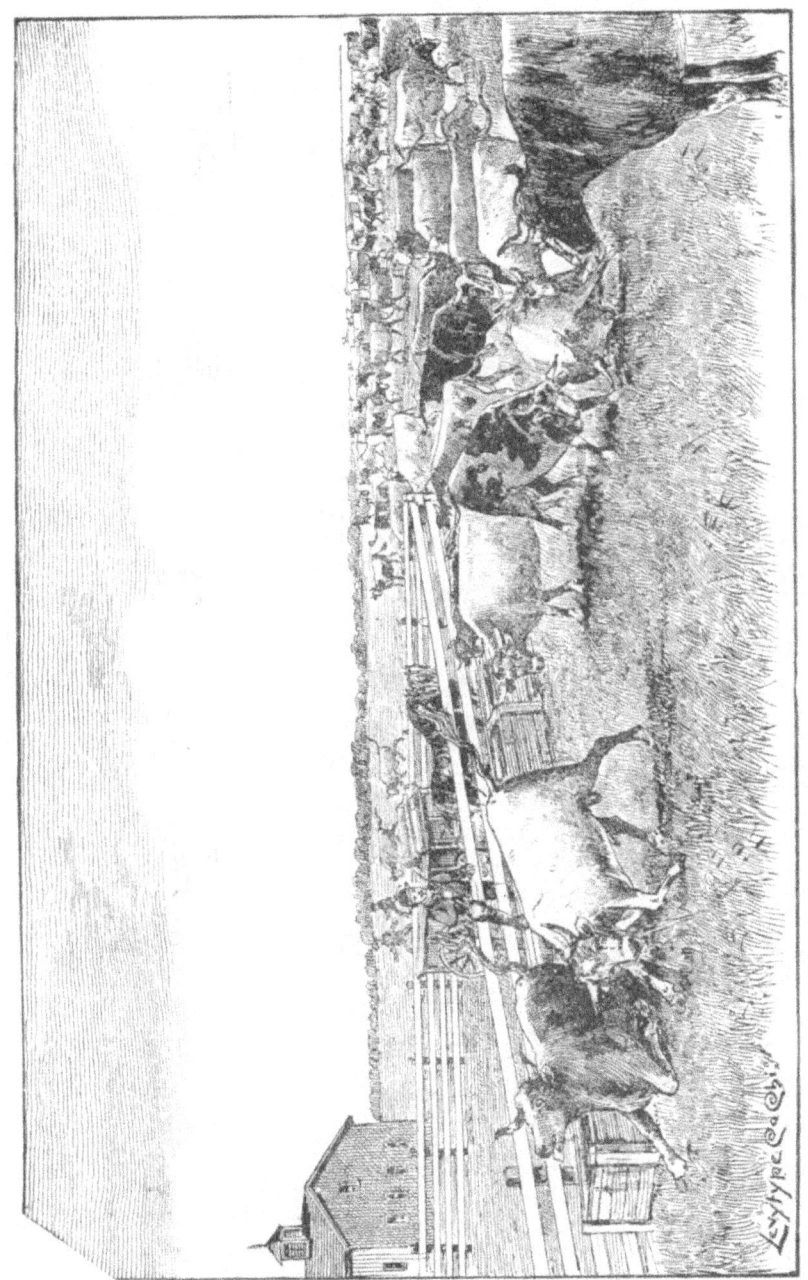

WHY "THE HORNS MUST GO."

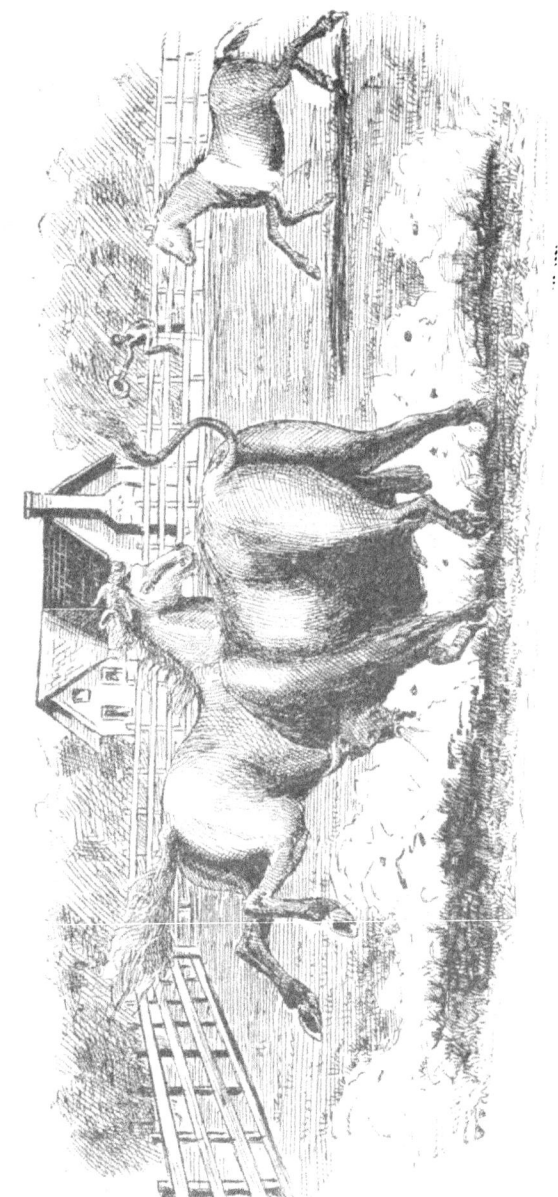

ANOTHER REASON WHY.

AND STILL ANOTHER.—"MILLIONS FOR DE-FENCE."

"GOOD-BYE HORNS."

HORNS.

Horns measure the constitutional vitality of the bovine. In nine-tenths of all North America, the ability of a stocker to survive the winter months is a heavy draft on the constitution of the animal. Given a pair of horns on a cow's head, twelve inches long from tip to base, (see Figs.), say from T to B, now it must be apparent that it requires more labor of the heart to supply blood for the animal economy if it must be pumped up a foot higher to "T" at two places than would be required to supply vitality if the horns were removed and the blood were to be supplied at B only, at the two bases of the horns. In other words, it requires food, blood and heart labor to supply warmth to the horns all the way up to the height of a foot at two points. It is well-known among physicians that in case of a cut artery on the arm the wound should be held higher than the heart, as it requires more heart labor to force the blood to the wound than it does if the wounded artery be allowed to hang below the level of heart. This statement is self-evident, and shows what little sense that man had who took his fellow from a mowing machine with his arm and artery cut and laid him down and ran for help instead of raising the arm and binding it to stop the flow of blood. Now, I repeat the horn measures, that is, it is a measure of the constitutional vitality of the bovine. It follows, therefore, that an animal with large horns must be able to stand the cold better than one with small horns. This is true, and accounts for the fact that the Hereford makes a better "rustler" than a "Short Horn." Hence an animal with no horns should be better able to withstand the cold than even a Hereford. True again as witness any of the polled breeds of cattle; any one who has had them along side the horns as I have, knows the fact. The horn lessens the constitutional power of the animal to withstand cold. Remove the horns from all animals in Texas, and not so many of them will succumb to the " Northers;" and dehorning cattle is one method by which the cow-boys can help to tide over a hard winter, and this statement will be recognized as a fact as soon as it has been tried.

The horn is a modification of the epidermis, or skin or hide, presenting the same structure whether in the nails of man, the claws of birds, the hoofs and horns of cattle, the spikes of the hedge hog, the plate of the armadillo, the whalebone of the whale, the quills of birds, the shell of the tortoise or the hair of the head. The horns of the stag and other deciduous (that is, those that are yearly shed) antlers, strictly speaking, are not horn, but they are true bone, and they are dropped off or shed by a process of absorption at the root like that by which dead bone is cast off after having been diseased, called necrosis. The structure of horn is a modification of cells, which becomes harder by drying, and becomes also more firmly adherent. These cells are arranged in regular layers, each indicating a period of growth. They are generally attached at the base to the corium or true skin (that is, the membrane which supports and gives life to the outer skin, and they are usually removed with the skin. They are secondary, growing and wearing away. They are liable to deformities by accident and are very beautiful in sections under the microscope. They appear much like a bundle of pressed hairs, as they are like hair in their composition and characteristics. In oxen and sheep and all hollow-horned ruminants (a ruminant is an animal which chews the cud) there is a central core of bone upon which the horns are moulded.

The horn received its name from the Latin word "cornu," which is a trumpet. It received its name also from the fact that it was formerly used as a drinking utensil, and hence the expression, "taking a horn." It is asserted that the cervus, that is the deer tribe, if dehorned, will lose their power of reproduction. This is an assertion, and I fail to find any proof of the fact, and I do not believe it to be true. This, however, I know to be true: the circulation of blood in the horn is of a secondary character and is principally capillary. I mean by that the blood runs up the periosteum or membrane in the same way or in a manner similar to that in which water runs up a handkerchief when suspended with an end in a basin of water; and hence the conclusion which I draw is this, that as the circulation of blood is wholly of a secondary character, the part is liable to be omitted and may be entirely bred off.

In all old animals the different bones of the head are liable to become united by the extension of ossification from one to the other, and while the parts may have been originally separate and independent, they are liable to become united into one solid mass or structure. This may account for the fact that some animals in the operation of dehorning will bleed slightly at the nose. The parts that in infancy were cartilage and separate have by process of ossification become somewhat shrunken and united so that the porous character of the cartilage has developed in an ossified form into direct openings, and it may seem to the person operating that there is a direct passage from the nose to the orifice where the horn is removed. There are various nasal bones which it is unnecessary to name here, but the bone to which I refer is technically called mesethmoid.

In the deer the horns are solid and bone-like in composition; that is, are not covered by hide or epidermis. We are so accustomed to talk of "a pair of horns" that we do not think it possible for a bovine to have more than two horns, and yet at the Stock Yards at Chicago there is now, or was for some years, a large mammoth·ox with three horns, the third growing from the center of the frontal bone at the apex or top, and equal or nearly as large as the side horns. This is simply another evidence of the fact claimed by the author, that the presence of horns or their absence is largely a question of breeding, and therefore a matter of choice on our part; and there are in existence now living specimens of four-horned goats and many-horned sheep.

The horn of the deer at first is soft, vascular and highly sensitive. They appear towards the end of spring. At this time the blood vessels surrounding the frontal eminences become large, and the budding horn grows with great rapidity. On attaining full growth a bar consisting of bony tubercles on the base of the horn is formed, and this, by pressing down, cuts off the blood vessels which supply nutriment to the antlers. The outer coating of the horn then begins to shrivel and all on account of the decreased circulation, and so the bone begins to die off, hastened by the deer rubbing the horn against the trees and rocks, and in process of time it sheds at the base, where it is connected with the frontal bone and drops to the ground; though in warm climates, as in India, the deer does not shed its horns annually, and it is said that in castrated animals the horns do not appear or are simply stubs.

Polled cattle are an artificial variety, which may be produced in any breed by selection and by persistent removal of the horns in early calfhood; and the polled cattle of the Galloway breed are known to have had horns as late as the middle of the last century; and by breeding with bulls with the shortest horns the Earl of Selkirk succeeded in removing these appendages. The same principle was adopted in producing the race of cattle we call Durham, or Shorthorn cattle; for they are the direct descendants of what were known in England as the Long Horn cattle, which are themselves descendants from Welsh and Highland cattle. Welsh and Highland cattle are the earliest and remotest breeds of cattle known to the British Isles, and date back in their ancestry to the Roman period.

The more one studies this subject of horns, the more he will find it true historically, as I know it to be true experimentally, that it is a question of breeding. This is abundantly shown in the ordinary Texas breed, and is also shown in the Hungarian breed of cattle, whose horns measure frequently five feet from tip to tip. These and the cattle of the Romans were believed to have been introduced by the Goths into Spain, and from Spain they were transported in turn to South America, and were the progenitors of the herds of wild cattle now roaming over the plains of South America, Mexico and Texas.

It was a consideration of these facts and other matters which came to my knowledge through my familiarity with cattle on my farm that led to my discovery of the art of dehorning cattle. A close examination of some of the illustrations given in this book will show the reader that at the base of the shell horn there is a visible jog or offset in the bone horn itself. I asked myself this question: If God has built the deer tribe, so that by a process of local strangulation at the base of the horn the animal in process of time sheds its own horns, the horns dying prior to being shed, and if in the construction of the bovine as compared with the deer horn, the striking difference seems to be that the former is covered with the shell horn, that is, that the hide of the animal is extended over the bone horn as the base or form, why may it not follow that if I remove the bone horn at the point where cervus sheds it naturally, and if at the same time I remove the shell horn at a point below the shell, why, I say, might it not follow that the bone horn itself would lack the power of elongation, and the shell horn be destroyed and lack the power of reproduction, and the orifice so contract as to leave the animal a well-shaped mulley. The thing was settled in my own mind in an instant. By a sort of intuition I jumped at this conclusion, put it in practice, and made the discovery, and I hope by giving my readers partially the mental process by which I arrived at my conclusion I shall be credited with honesty of purpose; for it is simply impossible that I should cover up anything in connection with this great question. I believe I am not boasting when I say that this is the greatest discovery of this age so far as the resulting benefits to the farmers are concerned; and I believe that I have made a second discovery, which, while not so great, is still to be of immense advantage to the farmers and cattle-men. That is my discovery of a successful cattle tag, by which the farmer shall be enabled to do away entirely with the barbarously severe practice of branding—a practice which, so far as the pain and suffering of the animal is concerned, is fourfold more severe than the practice of dehorning, and is, in my opinion, second only in severity to the practice of castration and spaying. If a gracious and munificent Providence shall spare my life for a few years, I know I shall see the cattle of this country wearing my brisket tags, and the practice of branding discarded, and if to this I may add for farm use a cheap and successful water or tank-heater (and I believe I have it already practically now), there will be but one thing more which I desire to do to round out my experience as a farmer; and I say it in all humility, without any wish to boast, that I believe that if my life shall be spared for five years, I will have ready for introduction among the farmers and grangers of this country a cheap and successful farm engine, which shall run without steam, without boiler, without danger of fire, without engineer to run it, and which shall be such a home institution that the women will employ it to do their washing and churning, to pump their water, and at the same time the farmer will use it to cut his feed, to pump water for his stock, to saw his wood, and to do his threshing. This may seem like boasting; but I shall be a terribly mistaken man if I do not in that time live to see it an accomplished fact. The first three—dehorning, the cattle tag, and the water heater—I know are accomplished facts; the last remains to have the finishing touches put to it, and I shall rejoice that thirteen years of blessed happy home life on the farm have been not wholly without resultant benefits to my brother farmers.

"THE REASON WHY."

It is now over two years since the famous trial of "The People of the State of Illinois *vs.* H. H. Haaff," for "Cruelty to Animals," which lasted for four cold January days, and attracted universal attention, and was decided for the defendant. More than a half-million of cattle have shed their horns since that trial, and fully three-fourths of the number came to judgment during the fall and winter of 1887-88. As if to add potency to the verdict, and with a saliant tinge of poetic justice in it, the president of the so-called "Humane Society" actually stood up at the last meeting of the so-called "National Humane Society," and publicly advocated the "dehorning of cattle" as a mercy—a humane practice—a kindness to man and beast; and while his opinion counts as the opinion of one man only, and that a man who knows little or nothing about it practically, still as a pointer—a straw—it shows the way the tide sets, and is not, in a retributive sense, "a bark lost."

"*The agricultural papers*" of the land have, during the past two years, almost without exception, advocated "dehorning" as a measure of practical value and money advantage to ranchmen, dairymen, feeders, and breeders alike. The doctrine advanced that "Dehorning on the Plains" could never become a practice, since the cows need their weapons to defend their calves from wolves and other wild animals is abundantly exploded by ranchmen, who write that it is little help to a calf to have a mother with horns, since two wolves usually go together, and one attends to the cow, while the other steals the calf. The losses from "screw-worms" from "frozen horns," "broken horns," and in "shipping and yarding horns," makes "dehorning" a *sine qua non* to every cowboy on the plains. Texas mulleys are by no means uncommon, and the horns are bound to go, on the big Western ranches, by the same law of economy that governs in the farm shed and pasture—to wit: the cattle are easier kept, are more docile; huddle in cold weather and keep warm; huddle in warm weather and keep off flies; handle as feeders easier and with economy, are more gentle and tractable as mothers; have no trouble at the water trough or tank; spend their spare time in cold weather in housing the manure under the shed; need to be fed as stockers but once a day, and that near noon, and finally when the hay is stacked in the field, will actually avoid the need of using pitchforks for spreading the hay, since all will surround the hay in bunches.

If more testimony is needed on the "reason why," it may be found in the appendix of letters from everywhere and from everybody giving personal experiences, even to the cowboy, who writes: "You may say that an old cow on the lift will get up twice as quick if dehorned."

"*The place where.*"—Cattle to be dehorned should, if possible, be confined in a cool yard, where they will be kept quiet or free from all excitement, as will be seen from reading the chapter on the *Horn and Head Bones.*

The action of the heart has much to do with a successful operation. All disturbing causes should be prevented or removed, if present. Loud noises, cracking whips, dogs, boisterous boys, strange animals, bulls, cackling geese and chickens are out of place during this operation. Anything that excites the animal must be avoided. Dehorn the old bull first, and turn him out into a pasture or a yard by himself. He will be in a quiet and a reflective mood while you are operating on the balance. Keep the cows and calves together while being dehorned, and for a week or so thereafter free from worry. If you do not, both will shrink. If you observe this direction, you will be surprised at how little they, either of them, care for the operation. It is such a comfort to cow and calf to leave them together, and, as far as possible, let them be for a week or so as they were prior to dehorning. So, too, with cattle of all kinds, feeders, stockers, cows and all. Old associations are soothing, new ones are exciting. To excite is to drive the blood to the matrix, and is liable to cause bleeding, especially in young cattle and calves, or those of a plethoric or full make-up of body, and hence it is always better that the cattle be not stuffed prior to dehorning.

DEHORNING CALVES.

In the discussion of this branch of the subject, reference is freely made to Fig. 1 and Fig. 2 of the parts of the head, and Figs. 25 and 26, which are representations of the "Outcutter," and Fig. 28, which represents the second tool used in dehorning calves, called the Gouge. Figs. 1 and 2 show the horn of a calf of an age too far advanced to use the calf instruments; but they illustrate what it is desired to show in connection with this branch of the subject, namely, at what time may the saw be profitably used on calves' heads. As a rule, to which there are necessarily variations, it may be said that four months is the time, and from three to four months is the time at which the outcutter and gouge cease to be valuable for the purpose of dehorning. The best time to dehorn the calf is at weaning time, or shortly before, and that is also the best time for performing the operation of castration. [See article on "Castration of Bulls."]

Now the question is asked—Are two tools really necessary to be used in dehorning calves? And the answer is emphatically, yes; and for the following-named reasons: As has been all along insisted in these pages, the operation of dehorning is more severe (do not quote me, gentle reader, as saying more cruel). I mean just what I say, "more severe." The operation of dehorning is more severe, or more bloody, if you please, than the operation at any age after calfhood. The reason for this is apparent, and, at the risk of repeating myself, I may refresh the reader's memory by saying it is because the parts are tender, for we must recollect the horn does not grow proportionately with the body of the brute, but much more rapidly, for the horn matures substantially at from two to three years of age; so at least this is true, that after three years of age the brute will take on many more pounds of avoirdupois in proportion than the horn will show in corresponding size, so that in removing the embryo horn from the calf, the question to be addressed is, how to remove the embryo and the cartilage underlying and leave the animal substantially not interfered with at any other part. Long experience has taught me that no single tool can be found (unless it should be a boring apparatus, and that would be horribly cruel), or can be used that will at once cut the hide, take out the embryo, and destroy the membrane underlying, by scraping the skull-bone so as to prevent reproduction, and at the same time leave a wound that will atrophy and so contract as to leave the head in an adult without cicatrix or scar that will show after the operation has been performed. I claim that no process of dehorning cattle is right which leaves such visible marks of the operation on the animal after healing has taken place; and if I don't teach and demonstrate a successful method of dehorning cattle from the calf of two days or two weeks up to the age of 30 years—which is the oldest age at which I have dehorned a brute—if, I say, I don't teach how to perform a perfect operation at any age and on any head, including as well the "nubbin on the doddie," the frail structure of the well-bred short-horn, the horn of the common farm-yard animal, or of the animals on the plains, which have alike been exposed to the rigors of open winter, and are so diseased that the membrane between the shell and bone horn has been destroyed—if I do not teach the reader of this book how to perfectly dehorn any of these animals, and at any age, then I have failed to accomplish the purpose for which this book was intended.

And now, reader, I say to you, that the operation of dehorning calves is the most peculiar, the most particular, and

DEHORNING CALVES.

the most severe operation in the whole category of dehorning cattle; in fact, it is a compound operation. First, the outcutter, shown in 25 and 26, must be used. The purpose for which this instrument is intended is to cut the hide through, not by striking down, not by a blow, but by a gentle pressure of the hand, and by turning the instrument on the calf's head. No man of ordinary intelligence will need to use this instrument on more than two horns, without discovering just how deep to cut. Turn the instrument firmly and fearlessly until you can feel that you are cutting bone, until you certainly know that at a depth of from a quarter to half an inch you have, by the twisting process, cut through the hide; through the underlying membrane and cartilage, and through the cartilage which has begun the process of ossification, and into the frontal bone, which is already getting hard. There is no need of cutting too deep, but be careful to cut deep enough. I have proceeded in this explanation far enough to say that you should throw the calf, and that his head should be firmly held on the ground or floor. I assume that every intelligent reader will understand that to dehorn a calf, or for that matter, any animal at all, the head is to be secured, and must be firmly held in a quiet position. You will need to use considerable pressure in cutting through the tough hide on the young calf's head—more pressure at the age of one month, than at the age of one or two days. In my judgment the time to dehorn calves, and I may also say to castrate calves, is when the horn-button first begins to protrude—that is, to show itself through the skin.

Having now used the outcutter marked O C in Figs. 25 and 26, lay it quickly down, and while the attendants still hold the calf and its head in a quiet position, use the gouge. By observing Fig. 28, you will notice that the jaws of the gouge are slightly elliptical, that is, they are not built on a circle, but the edges at the cutting point are a little down, so as to make them cut in using a little deeper all the time; in other words, in lifting out the embryo horn, the gouge as an instrument will not only lift, but it will also cut the membrane underneath, and will, at the same time, scrape the bone below the membrane. No gouge or other instrument for dehorning calves can be properly made unless substantially this form is adopted, and those men who are in the market with their "dehorning pincers" and their "dehorning knives," have simply stolen the ideas here presented from hearing me talk, or from articles written in the papers, or from my little book, "Haaff on Dehorning," and they will make an inglorious record for themselves among the farmers, from the simple fact that their tools will not fill the bill, or successfully perform the operation without first cutting the hide as above described. The first gouges were built for my own use (as all my tools were first built), but they have met with a continually increasing public demand. When a man begins to get from ten to a hundred letters daily from his brother farmers, demanding that he give them some sort of an instrument to successfully dehorn cattle or calves, he should do one of the two things, either simply say, I can't do it, or proceed with the other alternative, as I have done, and do it, and do it well. I will give a dollar a horn for all the calf horns that can be produced after the proper use of my gouge and outcutter, and I believe I have made this explanation of the operation so plain, that taken in connection with the cuts or illustrations, any boy ten years old can dehorn any calf.

I recommend in dairy districts or where stanchions are used that you throw the calf and put him in the manure trough, that is, the foot wide passage way at the rear of the stanchion for receiving the droppings—of course cleaning it and putting a little litter into it. Throw the calf, with the back in this manure trough, and then turn the head in the position seen in Fig. 25; let one attendant firmly hold the head between his knees with both hands, each hand holding an ear; let the operator stand in front, facing the calf; and let one or more attendants hold the legs; one good strong boy or a man straddle the calf, and holding the hind legs, one in each hand, down solid on the floor is enough; and there ought not to be less than three in dehorning calves. The attendant who holds the hind legs as well as the attendant who holds the head, can remain in their positions until the operator, having dehorned, proceeds at the other end to perform the operation of castration if desired.

Fig. 1 shows a calf's horn twice removed with the saw. This horn was too old for the gouge or outcutter. The first or upper cut shows the operation [for this was the actual horn sent me by the gentleman who learned the art of dehorning of me; and he learned it, by the way, after having sent me this horn]. The trouble with the upper cut is that the base of the embryo is not reached. The lower cut is right on each side and not quite right on the front. On either side you will perceive that more of the hide and natural matrix has been removed than at the front; but still enough of the horn and attaching parts have been removed to prevent its subsequent growth. H H shows the hair, and, by the way, the horn itself at this age is frequently hirsute. O is the orifice. The vascular character of the process has begun to show itself by ossification, while the part underlaying is still in a half cartilagenous condition. M M shows a part of the matrix, if we could see underneath we would know that the orifice O is extended through the entire length of the horn, and that in process of growth it does connect (as shown in horns of older cattle) with the frontal sinuses.

The perversity and mulish obstinacy of mankind was never more clearly demonstrated than in this matter of dehorning cattle. At this point of writing, while I was at my desk, in came the representative of a hardware establishment, and after inquiring my name and finding it, proceeded to read to me a letter from some correspondent—a dealer with his hardware firm—who writes about as follows: "One of our patrons wants a dehorning saw, if he can procure one without paying a fabulous price therefor." He has had bad luck in using a meat saw and wants one of Haaff's. He has discovered by experience that he cannot properly dehorn either his calves or older cattle with a common meat saw or a carpenter's stiff-back saw, and as a dernier resort he "comes to Haaff," still grumbling that he wants it if he can get it "without paying a fabulous price." Gentle reader, to dehorn his calves or older cattle properly, and make it a thing of uniform success, he has got to have "Haaff's tools," or something very similar. If I hadn't given the best years of my life to the services of the farmer without fee or reward, it might not be allowable in me to kick at the attitude assumed by some of these men, who "can saw off a horn as well as Haaff," some of them dealers who tell the farmer that any saw or other tool is as good for the purpose as Haaff's. I have borne with this sort of thing until forbearance has ceased to be a virtue, and I have small sympathy for a man who "has bad luck" in his dehorning under such circumstances.

A reference to Fig. 2 shows a horn of three or four, or possibly five months' growth, S S is the shell horn cut away purposely on one side. H B is the horn bone, or better, bone horn. The periostum, or membrane which produces the horn growth, is shown in this cut, but is not marked. The vascular or porous character of the young horn even at this age is shown in the several colored marks which appear in the center, but it will be seen by comparing Fig. 2, with Fig. 15, or other similar figures, how large the orifice in the horn becomes with age. In Fig. 2, the bone horn can be cut with a knife, something after the manner of a piece of soft wood, while in Fig. 15 no more impression can be made upon the bone horn with a knife than could be done upon a piece of "lignum vitæ." In the one case the calf horn is substantially useless as a weapon of attack or defense. A severe blow by the calf itself or by another upon the horn, would not only knock off the shell, but would give the calf a very sore head for a long time, and prevent subsequent growth; while in Figs. 15, or 19, or 21, or 22, which are figures of horns of older cattle, the same shock would produce little or no effect upon the horn or attaching parts. It is a rather singular fact that the character of the bone horn in a calf of the age of four or six months, like Fig. 2, will undergo the hardening process after being removed, so as to become substantially as hard as the horn of the adult brute. It is also a singular fact that where the calf or young animal is improperly dehorned, so that the stub horn grows as in the case of Fig. 4, or Figs. 5 and 17, etc., that the subsequent growth is very much less porous, is far more dense, and harder than the horn at any time during its first growth. Fig. 5 is the second growth of a stub horn. C C shows the point where the horn was first and improperly removed—possibly it is a little lower than the mark C C. The reader will notice that perhaps a quarter of an inch of horn was left in this case on the first dehorning. This cut is also an illustration of the dense character of the subsequent growth. Through C to the upper G almost the entire substance is shell horn. It seems like a pile, so to speak, of hardened hide, while at lower G the porous character of the bone horn is less distinct than in any specimens given. A small hole appears at G, but the balance of the bone horn in these specimens is so hard as to make it simply impossible to cut it with a knife, and it would polish like ivory.

A queer thing in connection with dehorning is shown in Fig. 4, at the points marked R R. R R is intended to show a ridge, and the ridge is plainly visible. This ridge, R R is true shell horn. It is as though the cut at the base at this point had been made low enough. The same thing appears in Fig. 17, at the point A; the same thing appears in Fig. 6, at the point A.

It will be seen that in the case of a young brute as well as in that of an old one, if the germ of the bone horn is properly removed the extension of the hide will largely continue so as to almost entirely cover the cicatrix, but if the germ of the bone horn or living process be not entirely removed the power of reproduction will be continued on one side, and where it ceases, as at R R in Fig. 4, or at A in Fig. 6, there the substance which oozes from the wound in dehorning will ossify, and partially or nearly cover the fractured surfaces and atrophied bone horn underneath.

This seems like a digression from the subject of dehorning calves and young cattle; but I deem this the proper place to illustrate the point, and if I repeat on this at another point I shall make no apology for so doing, except a desire to enable you to understand this process of the combined atrophy, contraction and covering of the parts where a horn was properly removed; for the combination of these three things was my discovery.

TO DEHORN OLDER CATTLE.

But to return. We now begin at the age of three or four months to dehorn cattle by the use of the saw. The question arises at the outset, Why not use at this age "nippers" or "shears," or some instrument like a chisel or otherwise, that will at one blow cut off the horn? The answer is this: The same experience that proves that the hot iron is not the thing to kill the horn in a calf demonstrates that shears, and nippers, and chisels cannot be successfully used to remove and kill the horn in older cattle. Burning will not destroy the germ in the calf's head without permanent danger to the calf; no more will clipping or nipping destroy the power of reproduction in the older brute, without great danger of injury to the animal.

Fig. 27 shows Haaff's saw frame. The pin or pinion at the front end and the slit at the rear are made for the purpose of receiving a very narrow and fine blade. I have had various blades made expressly for the purpose of dehorning cattle. I have tried all kinds and sizes, and the result of much experience is this blade: The Haaff blade is neither too soft nor too hard; neither too wide nor too narrow. It is now built so that a two-inch knife blade is furnished at the rear. This knife blade is

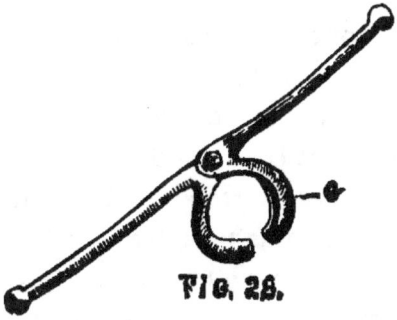
Fig. 28.

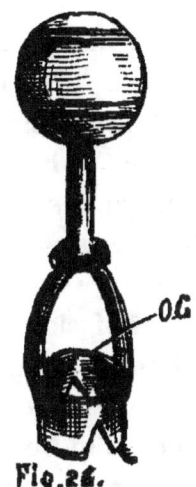
Fig. 26.

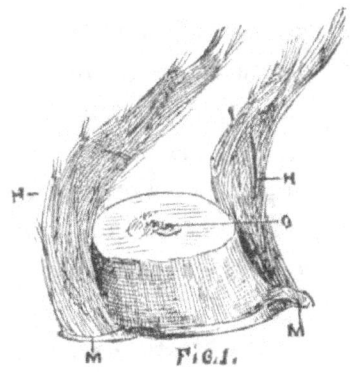
Fig. 1.

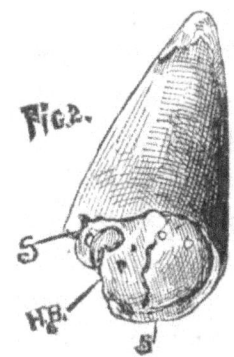
Fig. 2.

Fig. 25.

to be sharpened and used after the bone horn has been entirely cut off. Cut from the top down, as a rule, toward the ear. Cut hide and hair indiscriminately with the saw. Do not shave the hair before operating; move it back a little with the left hand, so that when the operation is performed the hair will fall over and cover the wound as much as possible. When you have cut off the bone horn so that it hangs by a loose figment of skin and flesh, which, by the way, is of course a part of the matrix, use the knife blade on the end, and by one dexterous stroke sever the hanging skin so as to leave as much and no more of a cut, proportionately, than there was at the beginning on the upper side. About a thousand people have written me to know why the pin or pinion at the end is not made square. In the hands of the new beginner—the man who has not yet learned to carefully and properly secure the head before beginning—there is some danger that between the struggles of the animal and the half-scared operation of the new beginner, the blade will be broken. These blades are therefore hung on a swivel at the end. If, however, the operator conceives that he runs no risk, it is a very easy matter to make the blade rigid at the end. Take the pin out; file it with a three-cornered file a little, and file the notch also in the shoulder where it slides. Shove it back to its place, and insert the end of a small nail or any small bit of iron to serve as a key. Before using the saw at all screw it up so tightly that it will sing when thumped with the finger-nail, and never dehorn with a loose saw-blade. I am free to say that now that my new mode of securing the brute is given as shown in Figs. 23, 24 and 30, the necessity of having the blade so that it will turn like a swivel and not break is substantially removed, as it is impossible for the animal to move in struggling. Still no harm can possibly follow, for the pinion can be made rigid at any time, if desired, as explained above.

Figs. 9 and 10 give the horn and upper skull bones of a yearling. I wish to call the reader's attention for a moment to these cuts. P P in Fig. 9 show the periosteum. H B is the bone horn. C C is the usual but improper place for cutting the horn on the yearling. C C, or as C G the artist has made it to appear, is the proper place. It will seem to the casual observer that there can be very little difference in the locality, but even a quarter of an inch on the end of a man's nose may make the difference between a pug nose and one that is not, and so this quarter of an inch is very important. By carefully observing Fig. 9 the reader will see that at the point G the vascular character of the bone horn has ceased; that it has substantially ceased or nearly so at C. Now C is next to or near to the ear, and it is the lowest point at which we can cut with safety in the case of any animal, and there is this peculiarity which I have found, there is less danger of a second growth at the base of the horn than either at the sides or the top. Why I do not know or pretend to say; but I know the fact. Now in the case of the yearling this peculiarity exists; a cut made at G in the bone not only "shapes the head," but it will also heal, that is, I mean the bone will fill up and cover the opening.

Now, if the reader will turn to the parts taken from the head of a three year old bull, Figs. 6 and 7, he will note, Fig. 7, the orifices G G are filled; so also the same thing is seen at O, in Fig. 6. The hole is filled, but at G G the cut is made below the filling. Now if the three-year-old (Figs. 6 and 7) had been cut at the point G in Fig. 9, the power of reproduction is so far gone in the case of the three-year-old that the filling will not take place, and a hole would remain in the frontal bone. Why? because the ossification in the three-year-old has become so dense as to have absorbed and destroyed the power of growth. It will thus be seen that the cut in the case of the yearling will be quite different from that of the three-year-old. In the three-year-old the saw should be used; placed so as to cut from C to C, while in the yearling the cut would be made more oblique or slanting, from C to G and never from C to C. That is, I mean to say that a cut from C to C in the head of a yearling will give a subsequent second growth of horn, while a cut from G to C will give a perfectly shaped mulley head of the yearling; and I mean to say that a cut from G to C on a three-year-old will be likely to leave the frontal bone with a permanent open orifice.

I cannot leave these Figs. 9 and 10 without calling the reader's attention to the awful suffering which the presence of the horn sometimes occasions the young animal. Any blow administered on the horn the reader can see will tend to separate the parts at the suture, which is shown in Fig. 9 at letter S. I think the reader can see by an inspection of these cuts that dehorning is a mercy even to the calf or yearling. As is seen at P in Fig, 9 and at G in Fig. 10, another bone intervenes between the frontal bone and the brain. The brain lies below P. At A in Fig. 9, as well as at P, is shown the parietal wall or bone which covers the brain. At S the suture is shown, and along the cross cut on either side of S the sinuses appear. From openings between the two bones at B in Fig. 9, the wall which sustains the frontal bone is shown. There are many of these walls or partitions, and I have not pretended to reproduce them all. Examine any skull yourself, and see how admirably adapted the parts are to the purposes for which they were intended. The bovine head is built with a second story; the horse and human with one story. Great strength was needed in the head of the bovine, and the two sets of bones produced a form of great strength. A blow on top of the head, Fig. 10 at F, which would kill a yearling colt would have little or no effect on a yearling calf. The cross bone at G binds the two parts of the frontal bone, holding them firmly together. One cannot but admire in the structure of this head the admirable mechanism of that God who has made all things suited to the purposes for which they are intended. Does the animal sustain a charge or blow on the head or horns, or does he make a charge with one horn, the second story, the cross bone and the partitions or walls between the frontal sinuses and the parietal bone make the head an almost perfect catapult or battering ram. Nothing in the shape of flesh and blood can withstand the onset. So true is this that Dr. Livingstone, in the annals of his African explorations, declares that a wild bull will more than match any able-bodied lion, and that he himself has seen a buffalo bull successfully keep at bay three or four full grown lions. I said that nothing in the shape of flesh and blood can withstand the onset, but it is a horse of another color when flesh and blood in the shape of a smart active boy hurls a stone or chunk of wood and strikes the horn as in Fig. 21, where the hand is placed. In this case, as I have elsewhere explained, the springing of the parts at the suture produces intense agony, not so much by knocking off the shell or bone horn, as by communication to the brain, caused by the unequal blow on the one side and the suddenness of the shock.

To return for a moment to Fig. 10. I have had the artist elongate the horn a little to show the location of the matrix, and how the bone horn gradually becomes harder, more dense and more like real bone at its base, until at D it is nothing but solid bone. I have also attempted to show in Fig. 10 where the matrix must be located. The reader will observe that at the base of the horn in Fig. 10, and also at the base of the horn in Fig. 19, there is a jog or offset. A A shows that jog. Cutting below the point A A, on this, the horn of a five-year-old, as will be seen by noting the wrinkles on the shell, would probably leave no stub horn, but it would be better to cut at M M, for it is found by practical experience that in the case of cattle of this age even it is sometimes better for the beginner to shave the edge of the matrix in the operation, and M M in this figure shows the matrix. The artist has also in Fig. 19 elongated the part somewhat to show where the matrix lies. Fig. 20 shows the same horn, and the same part, according to nature. In both Figs. 19 and 20 P shows the periosteum. A A is substantially the point where the shell ends. One danger in removing the horn of the adult bovine at A A is that some part of the shell may be left, and if left one of two things is sure to follow: either the stub will grow from the shell, or the head will matterate until that part of the shell which is left remaining rots away. In this connection it may be as well to say that sore heads in dehorning follow from a variety of causes. A remaining scrap of shell horn may produce it; a frozen horn is sure to produce it; a head diseased will likewise produce it, no matter from what cause the disease may come. Anything affecting the general condition of the animal is almost certain to develop a sore head after the process of dehorning; nor is it by any manner of means a bad result in its consequences to the animal. Many a man can testify that an animal that was "doing nothing;" making no growth; taking on no flesh; an animal that had been "off" for months, suddenly after the operation of dehorning has been performed, will begin to mend; will show a vigorous appetite, lively disposition and take on flesh at a rapid rate. In a farmers' convention at Madison, Wis., this year, there were several men who called my attention to their experience in this line. In the convention at Woodstock, Ill., where the subject of dehorning cattle occupied several hours, one gentleman gave an instance of a pet cow, an old animal, and so great a favorite that he refused to be present, and purposely went to town while she was being dehorned, and his son superintended the job. This old cow very soon after the operation gained in flesh, in spirit, in milk and in appetite; but there was a matterated head, and this need occasion the reader no alarm unless it has been done by some carelessness in operating. If you have allowed the animal to bruise its head by a sudden jam on the matrix and the adjoining parts either before or subsequent to the operation you will be sure to have a matterated head until nature restores the bruised part. The danger is in this case that the large veins and arteries at the base of the horn next to the ear may have been so bruised that the inner coating shall have lost its contractile force, and subsequent hemorrhage may follow, and consequent loss possibly of the animal itself by bleeding; but this has been treated of more fully in the article on bleeding. A ring of shell horn on the head is almost certain to be followed by suppuration. I have shown how bleeding necessarily follows in such a case, and where there is much bleeding there must be much suppuration to restore the parts. I could fill pages in recounting the experiences of different ones with sore heads. I heard of a man this winter out in Kansas, who lost four or five cattle out of one herd, and said he, " If this is dehorning we don't want any more of it in this country." My informant, who is himself an experienced dehorner, in writing me of this case says that the stub horns and the awful wounds made by the bungler who did this job was a sight calculated to sicken one, and disgust him with the operation, and he thinks it is a duty to the public to prevent the possibility of a recurrence of such proceedings; and I am writing this book, gentle reader, for that very purpose. Those four or five cattle which our unknown friend lost, and the damage to his herd of a hundred or more, which my friend says was an unwelcome sight, the cattle looking as though they had just passed through a severe winter—hollow, gaunt, wandering aimlessly around, without appetite—the loss, I say, to that one herd of cattle, would have furnished enough of these books to have made every farmer in the country a practical dehorner. Rest assured I will do what I can to prevent this kind of amateur performance. I don't believe a farmer can be found who will say that the cost of the book and proper tools is not a good investment in the case of every man in the land who owns cattle.

THE VARIOUS MODES OF SECURING THE ANIMAL.

The first thing to be done in dehorning an animal is to properly secure the animal, for no one can successfully dehorn cattle, unless the animal is quiet, and unless the parts next the horn especially are so as to be substantially immovable. No one can successfully dehorn a brute when its head is liable to be jerked either up and down, or sideways. After my discovery of a successful method of removing the horn, the next great question that presented itself was this—of securing the animal. I tried casting and tying with ropes. I tried lashing the head and neck between two trees. I tried tying the head to a post, with the frontal bone brought square up to the post, so that a horn on either side could be reached. I tried the stanchion, tying up as well as fastening the head and neck. All these plans are attended with great inconvenience, much annoyance to the brute and operator, and much of danger to both. The animal is sure to become terribly excited, and to thrash around in such a way as to become heated. The operator is sure to become nervous, in fact after a little, if he be not a man unusually well balanced, his temper is apt to be somewhat riled; in fact he gets something worse than "buck fever;" he gets a fever of mad, and is apt to damage himself and the brute too. No animal is likely to be properly dehorned while either man or brute are hot or excited. After a variety of experiments I ceased attempting to secure the body of the animal, and adopted my method of dehorning by the use of the stanchion. The loop, showing the rope with two rings in the cut marked Fig. 1, shows how the rope should be placed upon the neck. After being laid over the neck the rope is doubled, and the double portion passed through one of the rings. Then the head should be raised and drawn up tightly to the top rail of the stanchion. While one or two attendants draw on the rope, a third should raise the nose; then bind the rope again over the top rail and nose, slipping it a second time through the second ring. I have found by experience that two three-inch rings are more convenient than one large one. In the stanchion cut, marked Fig. 2, which is the same cut that was shown in "Haaff on Dehorning," only one small ring is shown, but, like a dozen other things, experience taught us how to improve on this method. The animal's head, in Fig. 2, should appear drawn up tightly to the top rail. So soon as we begin using the saw the animal will, in three cases out of four, fall over, and, if properly tied, will strike on the hip, resting on the hip and hanging part by the neck and head. The illustration given and marked as stanchion, Fig. 3, shows the cow after having fallen; but in this case, as in Fig. 2, the head is not right. In the one it is too low, and not drawn to place, while in the other the nose is too high, and the rope should appear drawn tightly around both nose and top rail. Some one or more persons have written me, wanting to know what holds the "movable 2 x 4 oak" and the catch above. A moment's consideration will show anyone that the top and bottom of the movable piece should work between two 2 x 6 pieces. The front, 2 x 6, both at the top and the bottom, is withdrawn, in order that the catch and the arrangement of the parts may appear in the cut. The space for the neck should be five inches in width, more or less, according to the size of the head and neck of the animal to be dehorned. If a bull, it will probably be even necessary to leave the movable bar clear down. The stanchion should be five feet in the clear, up and down, for the space where the neck of the animal is confined. That will give you a stanchion six feet from floor to top of top rail. For ordinary use this is a little higher than is necessary, but it is none too high for dehorning purposes. In either instance, great care should be taken that the legs of the animal, particularly the hind legs, be not caught in some way among the parts of the stanchion. To avoid this trouble, it has been my rule where I could do it, to have half a dozen stanchions or more together. By having so many cattle standing closely together (for two feet and a half in width is enough for almost any brute), the reader will see that all possibility of getting the legs mixed by twisting or turning in the stanchion is avoided. The artist has not been true to the fact in giving the location and appearance of the animal, but the cuts will do well enough with my explanation. Having removed the horns, or having bound the head and nose and removed one horn, it is sometimes necessary to turn and bind again and remove the other; but when both horns are off, the stanchion should be thrown open, and the animal invited to rise at once, and allowed to move out into the open yard.

I hope none of my brother farmers, however moderate their circumstances, or however small their ability as carpenters, will neglect the building of a suitable stanchion, one or more, according to the number of milk cows they have, for whether summer or winter, a stable for the cows for feeding and milking purposes is essential to attain the best results; and with the horns removed, it makes the matter of keeping cows much more pleasant in the stanchion or stable. The labor of building half a dozen stanchions is far less than one might suppose. Study the cuts and figure out the distances and lumber for yourselves. Anyone can do it in five minutes' time. Observe that the upright at the right hand is one twelve inch plank, cut to the shape or form shown in the figure. The wide part at the bottom is necessary, so that when the movable falls back, the animal, in putting its head into the stanchion, will put it in the right place. There is absolutely no waste at all in cutting these uprights where 2 x 12, twelve feet long, are used, with 2 x 6 stuff, twelve feet long.

It is now my pleasure to present to my readers one other mode of stanchion, that invented and used by Mr. E. P. C. Webster, of Marysville, Kansas. Brother Webster is a "dehorner from wayback," to use a slang phrase. He is one of those practical, everyday kind of men, who size a thing up for just what it is worth. His motto is substantially, I take it, the

THE STANCHION.

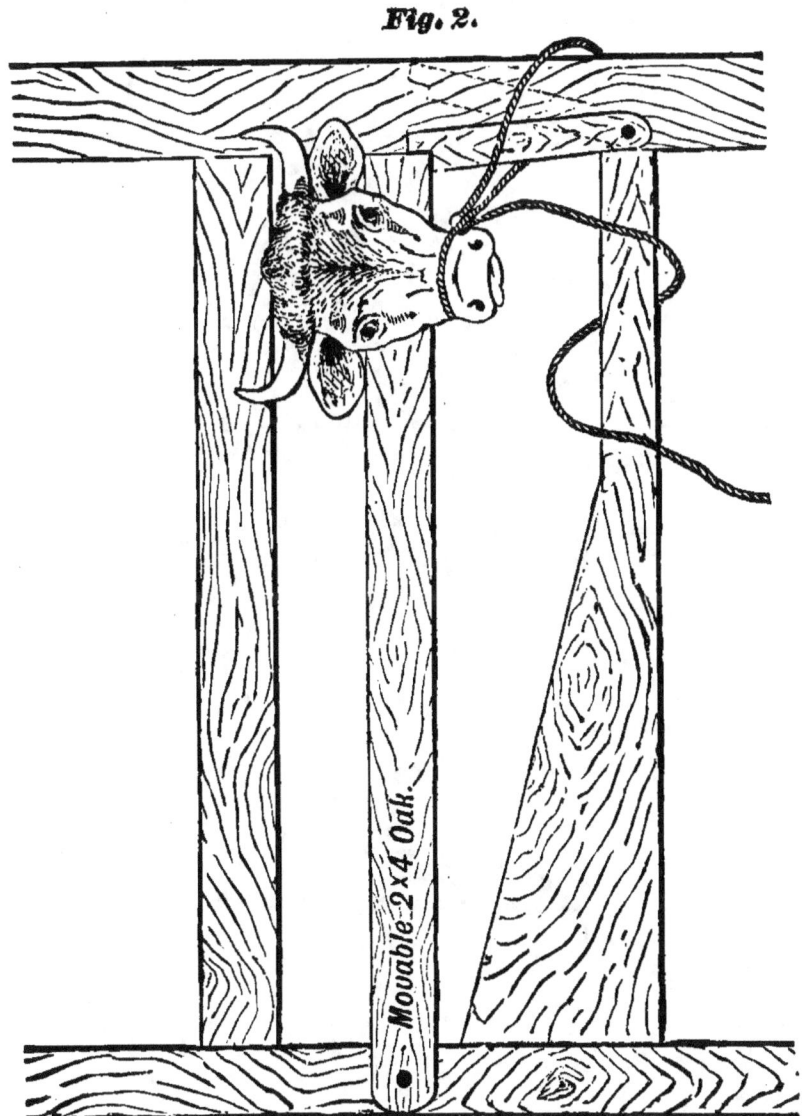

Fig. 2.

THE STANCHION.

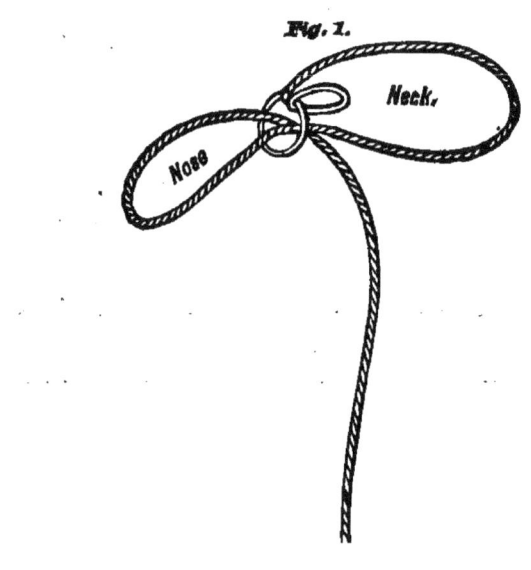

Shakspearian motto regarding the utility of anything presented, "Will it give me bread, or set a bone? No; then I'll none of it." Traveling as he does over a large portion of Kansas, dehorning cattle everywhere by the thousand, not only in Kansas, but in other states, he finds it necessary at every place to provide some secure, and at the same time cheap and easily constructed mode of holding the animal. He sends me a photographic view of one of his stanchions, which he built one morning this winter, and dehorned for a farmer some fifty odd head of cattle, and drove six miles afterwards to another point, and did all in the course of the forenoon, so he was on hand at noon for his dinner; and all this besides making

"THE WAY MR. WEBSTER DOES IT."

a similar trip in the morning to the farmer's place. A careful inspection of the cut discovers a twelve-inch board or plank at the bottom, securely nailed at the two ends to two upright posts, and of a board or plank at the top to correspond. If I readily understand him, the boards or planks at the bottom and at the top are double; in fact, I know they must be, because a catch is shown in the illustration which fastens the movable bar of the stanchion up against the animal's neck. It will be observed by examining this cut that Brother Webster has inserted several pegs in the upright to which the nose of the animal is directly attached. These pegs, he explains, are located at different heights to accommodate different sizes of cattle;

they are three hard wood pegs driven into an inch, or three-quarter inch hole, or holes, bored into the upright at convenient points. It will be observed, also, upon inspection, that the animal's neck is tightly bound to the upright on one side of the stanchion where the animal is held. Mr. W.'s idea is to put the animal in a substantially immovable position by the double bind, binding the nose to one upright and the neck to another. This gives the operator free access to both horns at one time, and obviates the difficulty experienced in using the stanchion as I had been accustomed to use it; for in my case it was frequently necessary to change the position of the neck, and turn it to the other side in order to get at the second horn. I would suppose the rope which appears in Mr. W.'s plan under the eye and along the cheek bone of the animal would be liable to irritate and injure the eye, but he says in practice this is not the case. It is certain that among the thousands of farms where there are no chutes and where the stanchion must be built, if at all, in a place like that shown in the cut, it is most probable that for places like this, Brother W.'s plan takes the cake, and will outrank all others. The same objection, however, pertains to this plan as to my own with the stanchion, and in fact to the plan of Mr. Richards of Iowa, which I will presently give—that is, getting the feet tangled in holes or open places, or partitions adjoining, and especially in twisting the neck. It seems to be an uncomfortable posture for an animal, and while it is only for a moment, yet I believe it ought to be done away with if possible, or convenient at least. It is only fair to add that Mr. Webster is not only a practical, but he is also a very successful dehorner of cattle, and that among the thousands and ten thousands of cattle that he has dehorned he has to loose his first animal, and this is more than I can say for myself.

The third method of securing the animal which I shall give is that of Mr. W. H. Richards, of Cresco, Iowa. It is shown in the accompanying cuts. This is a portable stanchion and chute combined. The parts are planks and boards, two by fours and four by fours, for posts and cross-bars. Three or four men can readily pick it up, throw it on to a common lumber wagon, and carry it anywhere. It is braced by iron rods, and to avoid the leg-tangling trouble above mentioned. Brother Richards has invented and added a wide belt, surcingle or strap, that passes under the animal and is drawn up by ropes, so that the weight of the animal rests partly on this strap or belly-band. The band, I believe, is about two feet wide; can be made of any strong canvas, as shown in the cut. This prevents the animal from lying down, and obviates the difficulty experienced in my own stanchion about getting the legs caught. It will be observed by inspecting the cut that the rope drawn over the animal's neck prevents its jumping out of the chute forward. It will be noticed also that there is a pulley attachment at the rear, by which the head and nose are drawn around and bound to one side of the portable chute. The pulley or roller attachment is added to furnish greater power, if needed. Mr. Richards has this chute patented. He goes everywhere, dehorning thousands of cattle, and with uniform success. He will give anyone the exact dimensions of his stanchion upon application by letter or otherwise. He backs his chute right up to the stable door, or a pair of bars in the yard, and the cattle are driven up, and, as one passes through the chute, he is caught and at once secured and dehorned, and all is done in less time than it takes me to tell it. A very small fee is added by Mr. Richards for the farm right to use his patent chute, and I can see, that in a great many cases, it is a most desirable tool to have; and after the dehorning is done for very many purposes on the farm, such a tool is well worth the two or three dollars that it will cost.

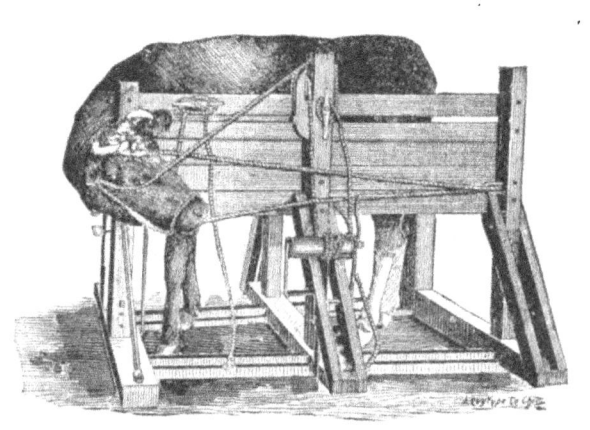

RICHARDS' DEHORNING CHUTE AND FRAME.

THE CHUTE.

It was not until the past summer that I was brought face to face with the absolute necessity of producing some other mode of securing the animal than that afforded by the use of the stanchion. The Hon. R. H. Whiting, of Illinois, has two sons located on large ranches in Kansas, and both of them are well stocked with hundreds of the very best of short horn cattle—cattle which have roamed wild over his ample thousands of acres for so many years that they have become virtually wild cattle; and they were worse than any wild cattle I ever saw. When it was understood by the neighbors that I was coming to Kansas to teach the Whitings the art of dehorning, a great many people in that locality were curious to see how I would manage to secure and hold those cattle. There were great twelve and fifteen hundred pound cows and imported bulls, and many of them so fierce that no man dared go on foot into the fields where they were, if likely to be near them.

There were the finest bullocks, and the finest calves, and the finest heifers it was ever my pleasure to see, coming off grass alone. I confess, having arrived on the ground and hearing the remarks that were made, and noticing the sly and grim looks of curiosity apparent among the cow-boys, and observing that the women folks of the establishment had found it convenient to make a visit to distant friends, taking the children with them, I confess I say, that I made up my mind that the outlook was rather bilious, and I didn't sleep much during the first night of my stay at the Whiting ranch. Before morning, however, the clouds had broken and the sky was unobscured. Mr. Whiting's entire ranch of many thousand acres and his yards were surrounded by stone walls six and one-half feet in height. Into one of the smallest o these yards were driven a hundred of these wild cattle. Previous to this we erected, with posts and by the use of 2x6s, two small yards containing about a dozen or twenty cattle each, separated from each other by cross-timbers of the same material. At the end of the yard there extended a chute, built of close material 1½ feet wide at the bottom and three feet wide at the top, giving the chute a V-appearance. We would rush the two yards full of wild cattle, then close the outside bars and the bars between the yards. Having the two yards of cattle together it tended to keep them more quiet and prevent desperate rushes. But we needed every inch of the six one-half feet in height, and there were some cattle who vaulted those fences in spite of us. My readers may guess that the situation is interesting when a bullock or an old cow will climb up the side of a chute six and one-half feet high to make you a friendly call. It must be borne in mind that the posts along that chute were only one and one-half feet apart and were held firmly together at the top by withes or a number of strands of wire. I would use wooden caps instead of wire, and have a good wide plank at the top to prevent the heads coming through. I found that the use of my Jewel and the plank at the side made a perfectly safe way of securing the wildest animal. We would run into this chute several cattle one after the other, putting poles between them as fast as they came up, to prevent the possibility of backing into the yard again, and when I came to repeat the operation at Ames, Neb., I built a chute long enough to hold a string of a dozen cattle one after the other. It will be understood, of course, that we had to use a block and tackle in order to draw these wild and fierce cattle up to the proper position at the end of the chute. The cross-bars which are to hold the cattle from going out should be so adjusted as to have a continuous bearing at the side of chute on the top and bottom of at least four feet besides the two feet needed to close the chute. This is done so that the bars can never drop down, but will be always in place, and can be slid into position with one finger.

Careful study of the figures will show how to build my chute, and I have steadily decreased the width on the bottom of the chute proper until I now use a single plank for the bottom. I have almost entirely discontinued the use of the stanchion in dehorning cattle. In using the chute I arrange holes on both sides, so as to shove two three-inch poles under the animal, about two feet up from the floor. The hind one should be omitted in case of cows heavy with calf.

This fourth and last method of securing cattle for dehorning is what I have termed " My New Method." I have told both Mr. Richards and Mr. Webster that while I would faithfully give, and be glad to give their plans an equal hearing in my book with my own (because I am writing this book for the benefit of the every-day farmer), yet I should be obliged to say that I believe my new mode " takes the cake." In the first place, all twisting and bending of the neck are done away with ; in the second place, the manner in which I construct my chute makes it, every way considered, a desirable piece of property on the farm whenever any bovine or animal of the horse kind shall need to be secured. My new mode is not patented; but a patent has been applied for, and the application has been made, in order to simply protect the every-day farmer, as well as myself, from the unjust attempt of outside parties, who are now, and have been, preying and poaching upon the labors and brains of both Messrs. Richards and Webster. I am not a little surprised when these men write me that all over the Northwest farmers will employ a lot of men to dehorn their cattle, who steal the brains of these men; appropriate their plans to themselves, and attempt to destroy their business by cutting under their prices. And here let me answer about one thousand inquiries like the following : " Will you also tell us in your reply to our letter what you consider to be a fair price for dehorning cattle." And I uniformly say from fifteen to twenty-five cents a head, according to number.

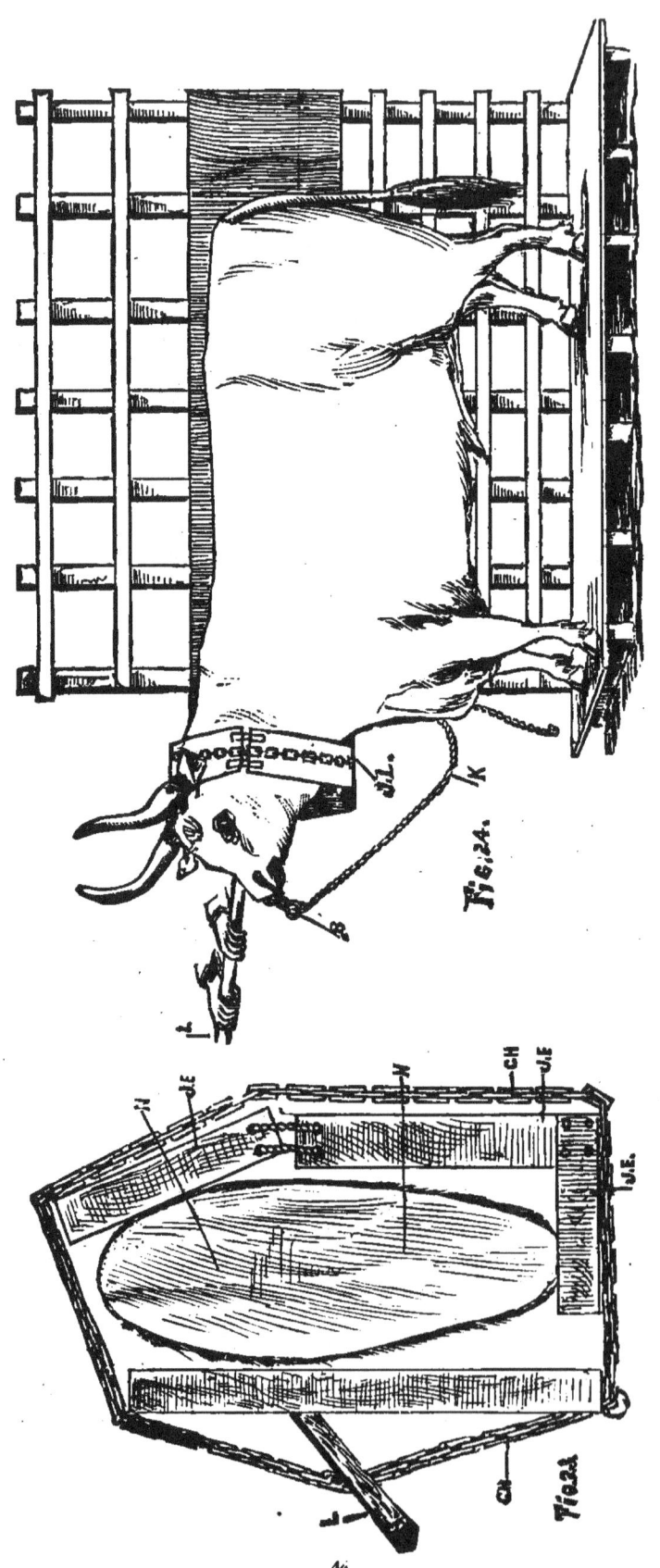

Fig. 46

Fig. 30.

Fig. 47.

location and trouble, all things being considered. In the case of herds numbering several hundreds it is quite likely that ten cents would be a fair price. But these mountebanks who go around the country dehorning cattle without knowing anything about it, without even having read my little pamphlet on the subject, who seize upon the carpenter's saw, or the butcher's meat saw; get out a few cards or posters, and style themselves, "John Smith, Dehorner;" going from place to place; charging five or six cents a head; leaving a lot of unsightly stubs; doing as a gentleman wrote me from Iowa (not either of the above) one man whom he knew did—killed two steers for a man the first thing in the morning. These men make the business of dehorning obnoxious, and bring it into great disrepute in certain localities. I am aware of the trouble, but do not know how to remedy it; and hence if the patent shall be granted me on my application I shall try to make it lively and interesting for some of the aforesaid fellows who "saw off horns."

I want "every man to be his own dehorner." I am writing this book for that purpose, and with that expectation. I feel, as the McHenry *Sentinel* puts it, just received and now on my table. It says: "Mr. Haaff addressed a large audience at our recent Fifth Congressional District Institute on the subject of dehorning cattle; and he made us all believe that dehorning has come to stay." It is for this purpose that I now offer my new mode, believing it for the reasons above stated, to be, all things considered, the mode of dehorning cattle that all ought to adopt.

An inspection of Figs. 23, 24 and 30 will give the reader a proper idea of what my new mode is. Fig. 30 shows the V-shaped rack or portable chute. On the right of the cow's head and neck will be seen in Figs. 23 and 24, and 30 the plank sticking out a foot in front of the chute, to which the neck is to be lashed. It is needless to say that the plank should be thoroughly spiked on to the uprights on one side or the other (I prefer the right side of the chute) and on the inside of the chute. The chute itself is composed of two sides solidly nailed together, the lower three feet on the inside being boarded up and down, as appears in part of Fig. 30; the up and down boarding is not shown in Fig. 24, but is purposely omitted in order that you may see the first part of the construction. In Fig. 24 or 30 the steer or cow is shown standing on one plank eight to twelve feet long, and ten or twelve inches wide; and herein lies the great improvement over the ordinary cattle chute. This chute is six and one-half feet high; inside, fourteen feet stuff should be used for uprights (cut in two), ten or twelve inches wide at the bottom, and three feet wide at the top. The loose pieces on top, shown in Fig. 30, are slipped over the top ends of the uprights, and go with pins to their places; or, as will be seen, the uprights can be extended, and the cross-bar placed so as to prevent the possibility of the ends being driven up. The lower part of the sides drop into the slots and are keyed. In Fig. 24 the ends should show four by four or four by six pieces on the ground, extending out a few inches beyond the plank, so as to be given plenty of chance to fasten the sides thereto. It will thus be seen that the two sides may in a moment be lifted away from the bottom, and the top pieces may be lifted off. The sides, bottom and side pieces may all be thrown into a wagon, and transported anywhere. Note one defect in Fig. 24: the plank on which the steer stands should not extend beyond the sides in front. This is remedied in Fig. 30, but in 30 the artist has failed to catch on to my instructions, for the two cross-bars at the front should show slots, into which they readily work, and uprights to them to prevent their being knocked off in the struggles of the animal. However, any intelligent person can remedy these little defects. Now, then, notice this—the advantages of this chute are: First, the narrow plank on which the animal stands; second, the close sides, preventing any possibility of injuring the feet or legs; third, its portable character; fourth, the plank, solid and firm, to which the animal can be securely fastened; and fifth, the fact that the V-shape of the chute prevents the possibility of the animal being cast in this chute. Of course the bar will be shoved in behind the animal as is shown in Fig. 30. There will be bars in front, and one or more underneath to prevent the animal lying down.

There remain now one or two important matters to be considered. Suppose the animal to be a wild, vicious bull or steer. He is driven into the chute. He tries and finds he cannot get out. As he jams himself forward an attendant with a cross-bar slips it in behind and the animal is caught. He can neither go forward nor backward. He is there. In an instant an attendant has thrown a rope noose over the horns, the other end of the rope is attached to the block and tackle. One or two men now draw on the rope in these pulleys. The animal is thus pulled forward, and his neck and breast are drawn against the cross-bars above and below. In an instant another attendant has thrown from the end of the plank, standing on the side opposite the plank, the arrangement shown in Figs. 23 and 24, which I call the "Jewel." A great many have asked me, "Why do you call such a looking thing as that a jewel." I answer, "Because it is a jewel." I consider it an immense discovery in connection not only with this practice of dehorning, but for use in any case in which the head or neck of a horse or bovine needs to be secured. The construction of the Jewel is plainly shown in Fig. 23, which gives us an end view. It is made of oak. Three pieces, 7 inches wide by 8 inches long are sufficient. The lower one under the throat should be in fact but 6 inches long. Seven inches is a good width for the neck from the chops down. Eight inches is about the right length up and down, for when the neck is secured and held in place by the pry, L, shown in cuts 23 and 24, the upper one of these pieces will fall over somewhat obliquely on top of the neck, so that the 16 inches of the two pieces will about accommodate the 12-inch plank to which the neck is to be secured. In this Fig. 23, the parts L is the 7-foot hand-spike; N N is the neck; J E the

THE CHUTE.

Jewel; Ch. Ch. is the chain. The chain should be about 3 or 3½ feet long. It should be fastened securely by a bolt to the bottom of the plank, as shown in Fig. 23. As I before stated, the attendant will stand on the side of the brute opposite the plank. He will catch the end of the chain. The chain, by the way, is stapled permanently and securely to the outside three pieces of oak which form the Jewel. The upper piece of plank is fastened to the lower piece by a couple pieces of rope, stapled on so that the upper piece cannot be turned on the chain. The attendant standing opposite grabs the chain above the top of the Jewel at the hook. No hook appears, but my reader will understand that the hook when caught on the other side of the chain will come somewhere about where the hand-spike appears. The attendant will pass this chain and Jewel up over the steer's neck; the attendant on the plank side of the steer catches the chain as it is passed to him, and by a quick motion fastens the chain hook on to the other part of the chain, drawing the steer's neck to the plank. The attendant on the plank side then quickly slips in a 6 or 7-foot pole or hand-spike, and drawing it out as will be seen in the cut, he fastens the neck of the steer so tightly to the plank that it is impossible for the strongest or largest bovine in existence to move laterally. At the same moment that this is done the operator should slip the bull-leader into the nose, and wind the rope attached thereto around the post at the bottom, or let the attendant on the other side do it; and grabbing the steer by the horns, put his knee on top of the animal's nose, and bring the nose and the head of the animal down so that the rope attached to the bull-leader will be drawn taut. The reader will now perceive that this animal can neither move his head or neck horizontally or perpendicularly. His head is immovable either up or down, or from one side to the other. In other words, we have at last discovered a way of properly and perfectly securing an animal for the purpose of dehorning him, and it makes little difference whether it is a 6-year-old bull, or a yearling heifer that is to be dehorned. A little more help may be required in one case than in the other, but the operation is the same, and is attended with no danger. Dehorning by this method is simple, easy and perfect. Dehorning by any other method, whether by casting the animal as the cow-boy on the plains does; whether by the use of the stanchion or otherwise, is terribly hard work, and no man ought to be in receipt of less than ten dollars a day who has to wrench, and strain, and draw, and twist, and turn, and lift and struggle, to get the animal's head into a position by the use of a rope or ropes in securing. In this case the slide bars in front work easy; the bar from behind is easy work; the two pulleys and ropes make the hauling of the animal easy work; the throwing of the Jewel over the neck is easy work; the catching of the hook on to the chain itself by the other attendant is easy work; the manipulating of the pry or hand-spike by one or more attendants, according to the size of the animal, is easy work; the pushing of the nose, and binding it by the use of the bull-leader and other rope, is easy work; and finally, when the operator has cast the rope off the horns, the use of Haaff's dehorning saw on the head of any animal that I have ever seen is comparatively easy work. The horns are free; so is the operator. The head cannot move. The operator can cut, and he will cut, if he understands his business, so as to make a perfect mulley without bleeding or danger of loss or damage, and with no possibility of having a stub horn afterwards if he has studied this book, and learned how to dehorn an animal at all. The time will come, and it will not be many years either, when the question will be asked of every man who attempts to dehorn cattle, "Have you studied and do you understand the modus operandi given in 'Haaff's Practical Dehorner?'" I say to you, my reader, and I say to everyone, you may call it egotism if you choose, there is but one true way to dehorn cattle, and that is the method here and now for the first time given to the public. I would not be afraid to guarantee ten thousand farmers and cattle men against possibility of loss or injury if they adopted this mode of handling their cattle during the operation. In fact, there is but one possibility of injury, and I am not going to cover that up, but I am going to tell it flat-footed and plain. After the Jewel is thrown over the neck, and the hand-spike has been placed and the chain drawn up, the bar on which the neck rests should be drawn forward—in fact, both the lower bars and the upper too, for that matter, may be drawn directly. This prevents the possibility of choking; and choking is the only way in which the animal can be injured in the Haaff chute. After the horns have been removed the nose rope should be loosened at the bottom, the bull-leader slipped from the nose and thrown to the ground at the side of the chute, where it can be picked up in a moment when the next animal is to be operated upon. Then the hand-spike should be quickly removed, and at the same instant the chain unhooked and the Jewel thrown from the neck. Of course it will hang at the bottom where it is bolted to the plank and will be ready for the next animal. The top bar and the lower bar need not be drawn until the Jewel is unloosed, but a very expert man will perform both these final operations in such an exceedingly small second of time as to appear to be done simultaneously; and the animal, which has not been confined during the whole operation one-sixteenth as long a time as it takes me to explain all this, will go out with a blat, kick and jump, and go on to his companions to all appearances as if nothing had happened to him. He will not begin to make the fuss that he would if he were branded; neither will the head be as sore, or so long sore as the spot where the cruel hot iron was placed. This Jewel and plank and chute leave nothing to be desired in the way of an apparatus for securing animals for the purpose of dehorning.

I ought not to drop this branch of the subject without saying to my readers: Brother Farmers, this chute will cost you possibly five dollars. It is as essential an apparatus to every well-regulated farm yard as a halter is a necessity for securing a horse in the stable. Every time you want to castrate a bull;

every time you have an unruly cow to be milked; every time you have a cut or wound on any quadruped to be sewed up or doctored, such a chute as this is the place to secure the animal. You know it. Why not stop and build it now? It may take you and your boy a day to do it; it ought not to take longer. Suppose it takes a day; what of it? Build it, and build it well. The more you use it the more you will like it; until every time it becomes necessary to secure anything from an old ram to a kicking horse, it will be a by-word on the farm, and you will all say, "Run him into the Haaff chute."

Figs. 8 and 16 are horns which were badly diseased. The casual observer would find it difficult to determine the difference in the disease; but Fig. 8 is a case of broken horn in which not only the shell was broken, but the bone horn itself was also broken, and the periosteum between torn from its place. In this case the reader can see that unless the animal is in a healthy condition and the weather very favorable it must suffer for a long time before the stub horn remaining would heal. The attempt of the periosteum to heal where it is torn from between the shell horn and bone horn fully illustrates what I have claimed as the certainty of much loss of blood where the horn is removed so as to leave a stub. In this Fig. S is the shell; P P the periosteum; O the orifice which is connected with the frontal sinuses. The stub below was about two inches long and the animal somewhat matured; probably four years old.

In Fig. 16 we have a case of destruction of the horn by simple freezing of the parts, and in both these cases there will inevitably be a partial destruction of the bone horn by reason of its slaking or dissolving; in fact, it seems to be the case with every diseased horn that the bone horn is likely to become more or less dissolved, so it makes little difference what the character of the disease may be, the result in every case is substantially the same, and the animal will do nothing until the diseased part is removed or healing takes place by natural process after a long time. Fig. 18 shows the horn removed from a five-year-old Texas steer. A comparison of this figure with Fig. 22, which is the horn of a seven-year-old imported Hereford bull, shows the back cut some times necessary to be made in dehorning. A moment's reference to Fig. 21 will show the difficulty of removing the horns from some cattle. Think of attempting to remove the horn from the brute's head in Fig. 21 with a stiff back saw, or with an ordinary meat saw either. The horn might be cut at A, but to cut it at C C with either of these tools would be an impossibility; and yet unless removed at C C, this brute, which bears the head of a three year old animal, is sure to have a stub horn. I have removed the hair to show the matrix, and where it should be cut on the head. The lower mark C is not quite far enough in. It should appear to be in nearly to the point of upper C. extended. To cut this horn off at A would be simply to give this animal a very sore head—probably for a long time, with very much hemorrhage at the time of the operation, and with a stub that in two years would stand out probably two inches long, and simply spoil the appearance of the animal, and detract largely from his value in selling. Now, by the use of my saw I can remove this animal's horn so that when removed it will have the appearance of the horns in Figs. 18 and 22, and so that in healing the part will atrophy, and the hair will presently fall over the wound; the bone horn will fill, and in six months' time, or as soon as a new crop of hair makes its appearance, the scar or cicatrix cannot be seen with the naked eye except upon very close examination. The Haaff saw will cut down into the head a little obliquely, and then the back cut to be made between the horn and the ear will strike the first cut at such a point as to destroy the horn and enable the hole to properly fill.

I wish just for one moment to call the reader's attention to Fig. 6. The difference in the two methods of dehorning named is plainly shown in Fig. 6. Facing the reader is a point, apparently a half inch or so, which was properly dehorned. It will be seen how the healing process has carried the shell horn in, while for three-quarters the way around the shell horn has begun to grow out already. The reader can see how the animal would probably look in Fig. 21, with one horn dehorned after the proper manner shown in Fig. 6, and the other dehorned improperly.

I wish to call the reader's attention in this connection to Fig. 3. This figure shows the horn of a two-year-old properly removed. S shows the end or termination of the shell horn; M M is part of the matrix removed in the operation of dehorning; O O are the orifices opening into the frontal sinuses of the head, and in this case a bar is shown between the two. I call attention to this horn for the purpose of showing how differently the shell horn appears in some cases than it does in others. It will be observed that the shell in this case does not extend to the base of the head, and it is for this reason that I have so persistently urged upon my pupils the necessity of making a different cut in older calves, yearlings, and some two-year-olds from that made in the case of older cattle. I call it "shaping the head." The power or capacity of nature for "pulling in" or healing the bone horn seems to alter or change after the brute has become two years old, and the reader must exercise some judgment in these cases, or he will be likely to have some stub horns.

Fig. 20 shows the formation of the upper part or parts of the skull bone connected with the horn, and shows also the shell horn as looked at from the inside. In this figure B marks the location of the brain; P W stands for parietal wall, F F are frontal sinuses; F B is the frontal bone; P the periostum and matrix; S is the point where the shell bone begins, but it does not appear in this cut.

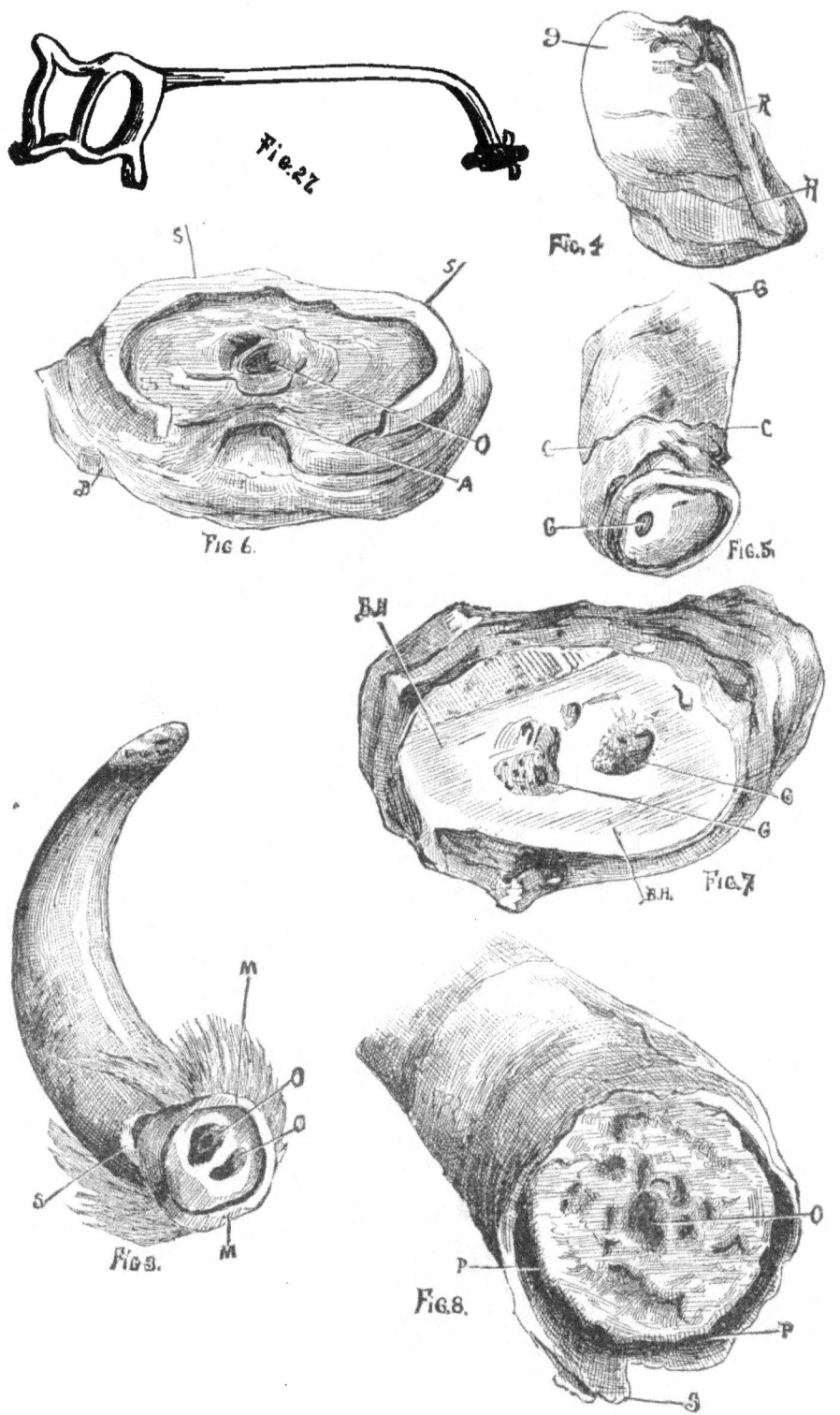

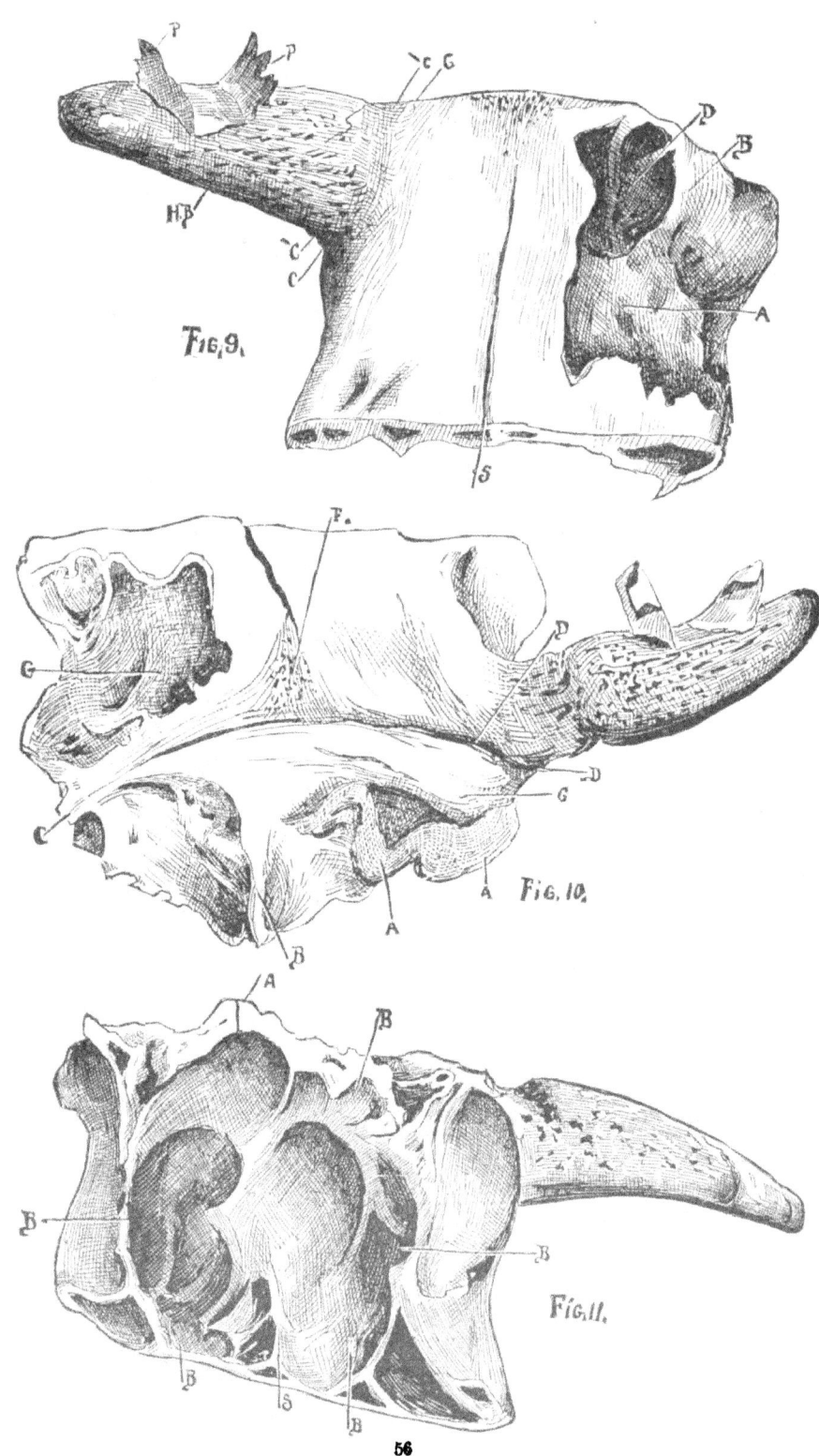

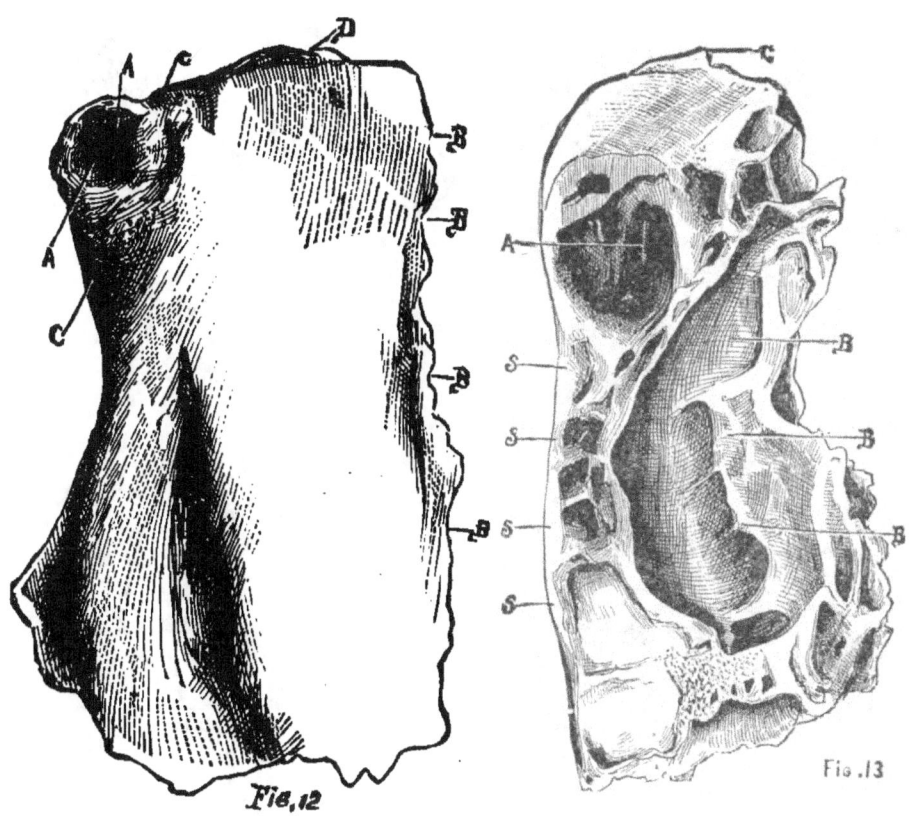

Fig. 12

Fig. 13

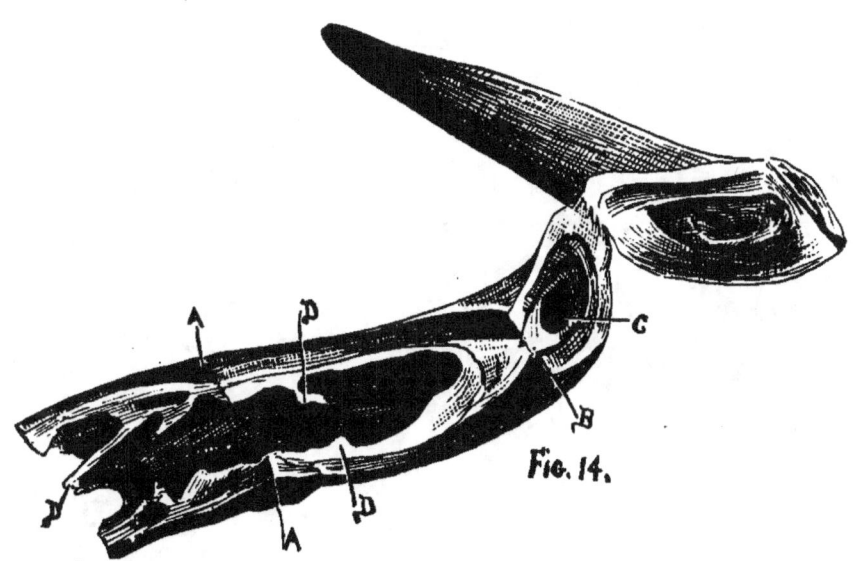

Fig. 14.

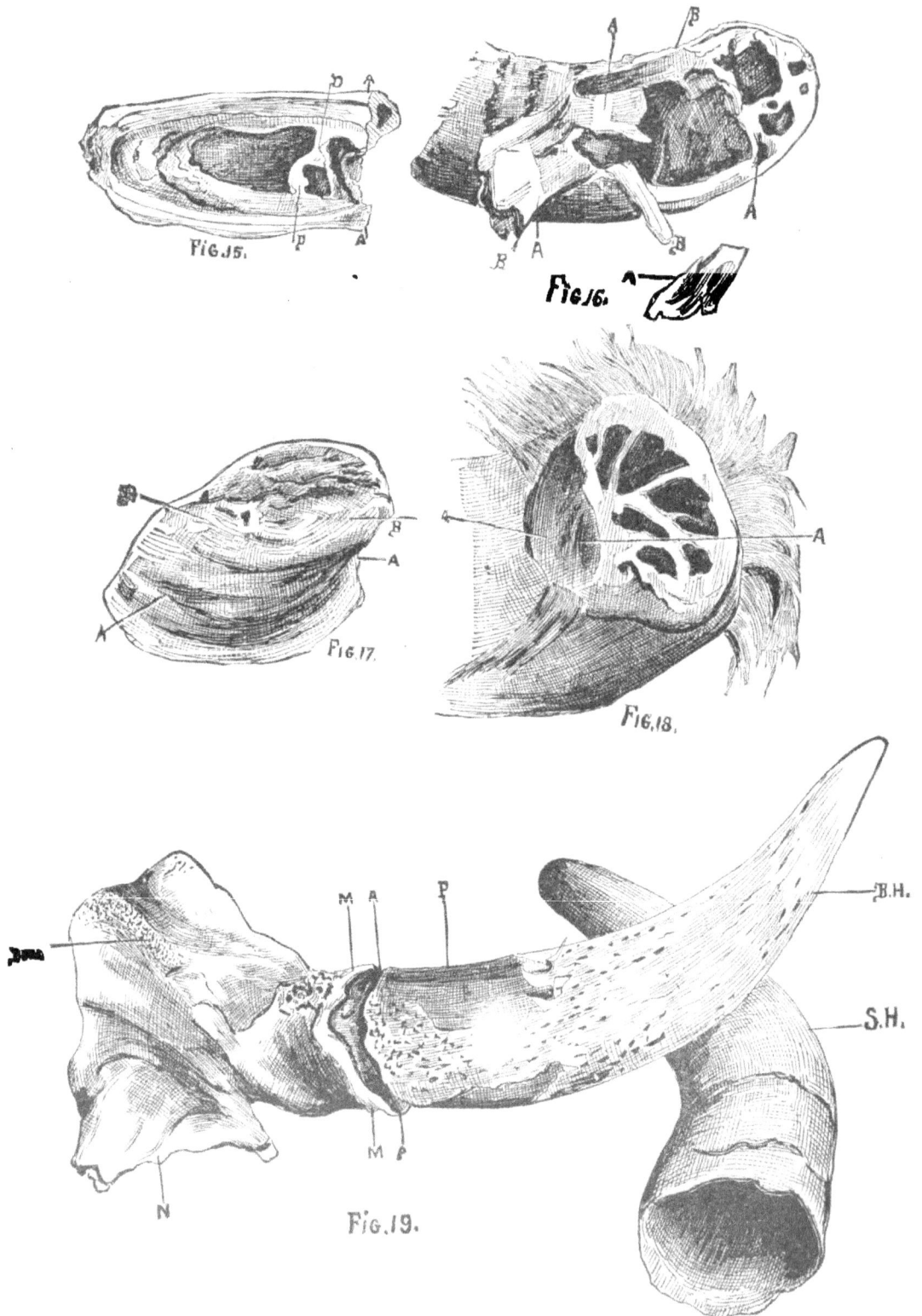

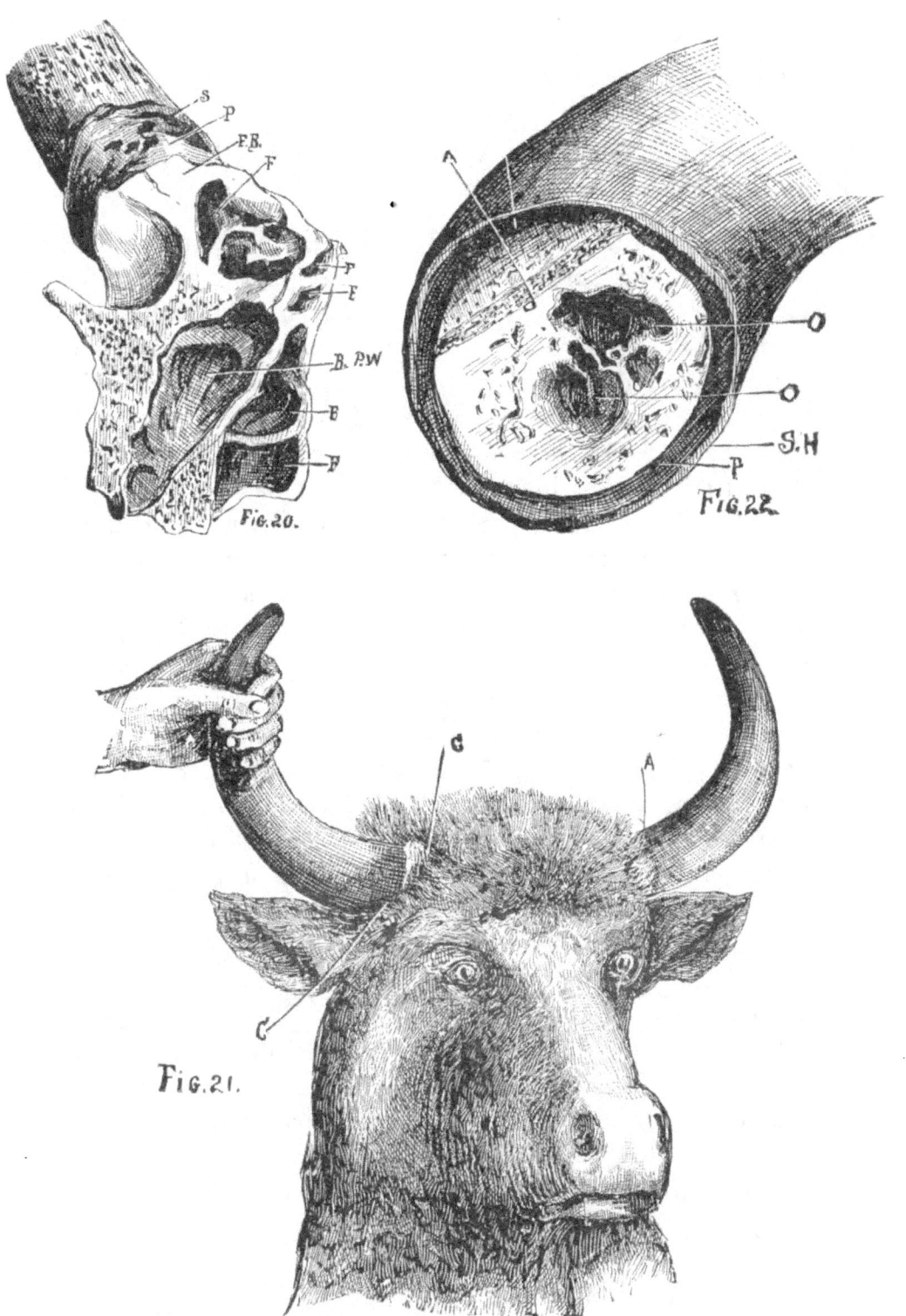

ATTITUDE OF THE PRESS AND PUBLIC SOCIETIES.

Up to the great trial the notice of dehorning cattle, so far as the press was concerned, was confined to the agricultural papers of the land, and foremost among these should always be named the *Western Rural*. The Hon. James Ward Wood, the real editorial writer of that paper, was personally inclined to wage a battle in my favor, but the proprietor, the Hon. Milton George (who is, by the way, not only a personal friend, but is one of the slickest business managers I have ever known) would not allow a positive public indorsement of the practice, but yet directed that the columns of his paper should be thrown open to the discussion of the question of dehorning, and once when the author had sent a particularly strong article, Mr. George was careful to explain that his position was a position of inquiry, and this was all that the author asked of any paper, and this is all that any person ought to ask for any new theory whatever. The motto of the author toward the press has been that of the boy to his father who was about to give him a licking. Said he, "Dad strike, but hear." The *Farmers Review*, of Chicago, and its able leaders, Messrs. Chandler and Gibbs, opened the columns of their paper to the discussion of the merits of dehorning, and, having thoroughly understood what was claimed for this practice, they have since advocated its adoption by their readers. The *Review* is entitled to credit for forbearance and consideration; it is an able paper. The *Breeders Gazette*, standing as it does at the very head of all the first-class strictly stock journals, has opened its columns to the author, and given by all odds the best account, through its Mr. Goodwin, of the convention of February, 1887, at Madison, Wis., at which time between 500 and 1,000 farmers, legislators and others were present, and the session on dehorning continued from 2 P. M. until 6:30 P. M.

This was followed by a terse and yet satisfactory account of the convention at the same place, in February, 1888, at which it was estimated that well nigh 1,000 farmers were present to hear this matter discussed and by a rising vote 66 men, according to the *Gazette*, voted that they had tried dehorning cattle according to Mr. Haaff's plan, and that they approved the practice and there were no negative votes. As the author has insisted that the Short Horn and Hereford men will have to adopt the practice of dehorning in self-defence, it is fair to presume that the position of the editor of this paper, Mr. Sanders, is substantially that of two previous papers mentioned, to-wit: "Let the discussion proceed and let the Short Horn and Hereford men decide it for themselves," and this is all the author asks. The truth of the matter is that these breeds of cattle are going to be crossed with the polled breeds of cattle. The author has called the attention of the Short Horn and Hereford men to this matter before, and he now reiterates his previous statement: You Short Horn and Hereford gentlemen will adopt the practice of dehorning cattle, or you will find that your customers will have left you; for the farmers of this country are going to use polled bulls in the immediate future, unless they can be satisfied that we can have polled races of Short Horn and Hereford cattle. I am free to say that it is my belief that before two years have elapsed the *Breeder's Gazette* will come to the front and acknowledge the truth of my proposition, and will simply say to the Short Horn men as I have done, "Gentlemen, there is no use in attempting to fight what is decreed by fate." "The horns must go," and your breeds of cattle must accept one or the other horn of this dilemma; either dehorn your cattle and help Mr. Haaff and the other advocates of the practice in their attempt to develop a race of hornless Durham and Hereford cattle, or else open your eyes to the fact that ninety per cent. of all the cattle men of this continent, from the long-horned Texas Racer to the Butter Jersey are going to adopt dehorning as a practice, and grow nothing but polled cattle. I know that some of the best cattle men in the ranks of that class of men who have made the *Breeder's Gazette* what it is, the leading exponent of the high bred cattle trade in these United States, have already adopted the practice, and are publicly advising their friends to do the same. I used to class the *Breeder's Gazette* among the Scribes and Pharisees, and I was not a little surprised to find that paper give place to such men as its associate editor, Mr. Goodwin, Prof. Henry and others, who are open advocates of the practice. It is refreshing to know that the number of Scribes and Pharisees is so beautifully small and growing less. I have never yet heard that the *National Livestock Journal* is a friend to the practice—in fact, I have somewhere a letter from its secretary which would seem to indicate that that journal is on the off side; I guess the *Country Gentleman* don't know yet whether it is afoot or ahorseback on this question. So far as the *Jersey Bulletin* and Hoard's *Dairyman* are concerned, they are the only Sadducees that I know of. There is a paper called the *Michigan Farmer*, which is said to oppose dehorning; but I do not characterize this paper, for I don't know that it is as yet sufficiently acquainted with the practice to speak understandingly. The *Ohio Farmer* has been kindly disposed and seemingly willing that the discussion should occupy a fair place in its columns. In New York and New Jersey *The Rural New Yorker* and *Orange County Farmer* have occupied during the year about the same position of active investigation as that occupied by the *Western Rural* and *Farmer's Review*. A year ago the *New England Farmer* and the *Massachusetts Ploughman* turned smiling faces towards the author, and have been willing to aid in the investigation; both editor Cheever, of the former, and editor Noyes, of the latter, have urged the author to come to Boston and preside at one of their farmers' conventions, assuring him that his presence at such a meeting would do much toward starting the dehorning boom in New England. The *Farm and Home*, of Springfield, Mass., and the *Farm and Fireside*, of Springfield, Ohio, with their enormous lists of

ATTITUDE OF THE PRESS AND PUBLIC SOCIETIES. 63

subscribers, have also to be numbered among this year's advocates of the practice. Turning to the extreme west, the *Livestock Indicator*, of Kansas City, has come out boldly, and editorially declares that "Horns Must Go;" while the Iowa *Homestead* of Des Moines, Iowa, *Farm, Stock and Home*, of Minnesota, among the best papers of the Northwest, are open advocates of the practice. It may seem invidious to distinguish where so many are willing to teach the people; but I deem it proper to mention those papers, nothing detracting or abating from the good will I bear all the rest, and which I hope they bear me, for the sake of the cause. The weekly metropolitan papers like the *Tribune* and *Inter Ocean*, of Chicago, *Plaindealer*, of Ohio, *Democrat*, of St. Louis, *World* of New York city, and many others have discussed the subject during the last year because they have observed that it was a subject of great interest to the farming population. If, as I believe, a half a million of cattle have been dehorned this year, where so many papers are seeking to teach the people it is safe to say that a million and a half of cattle will lose their horns the next year, provided always we shall be able to so instruct the owners that they can perform the operation properly. The *Dairy World*, of Chicago, a young yet very enterprising journal, whose editor, Mr. Birch, has been most liberal in his offer of premiums at our western dairy shows, and who has been personally of service to the cause among the dairymen has opened the columns of his paper to the advocacy of this practice, so that Chicago has come to be a sort of head center of the friends of dehorning.

I have had occasion to refer to the incongruous course taken by the State Board of Agriculture of Illinois in reference to the subject of dehorning cattle, and I have lived to see the members of that Board individually eat their own words. I have taken occasion to contrast their mulish obstinacy in refusing, at my request, to even give the subject of dehorning their undivided attention for the space of ten minutes, while at the same session of their Board they gave

Armour's cotton seed oil and oleomargarine men over an hour. They have lived to see that they lost an opportunity. They have come to know that the discussion of this subject fills the largest room in any interior town at the holding of their institutes, and they fully understand that the hour at which the subject of dehorning cattle comes up marks the biggest crowd of that series of meetings. All this is the logic of events, and from a manner dictatorial and full of spleen, I am happy to say they are now inclined to do the subject justice and give it a fair hearing.

The State Board of Agriculture of Wisconsin adopted a far different course. At their session at Madison, in February, 1887, there were not more than two of their officers who stood for dehorning. Professor Henry had tried it, and did not hesitate to say that it was worthy the trouble of investigation, and ought to receive careful attention and proper consideration. The professor is known as a very conservative man; a man of splendid natural abilities; of great powers of reasoning and demonstration; clear as a bell in his ideas, and choice in his language, he is at the same time slow in coming to conclusions, and he is very properly looked up to by not only the Board of Agriculture of his own state and the entire round of farming population as the head center of all that pertains to the cause of advanced agriculture in that state, but he is fairly allotted the front place to-day as a national experimental teacher, and the national representative of the agricultural interests of the land, standing, as Gladstone himself does, head and shoulders above his peers, so, too, Professor Henry, while yet dehorning was in its infancy, without having heard me lecture, without having seen me operate, yet stood squarely on the issue, facing down all opposition, and declaring that each for himself should see, hear and decide after a full and thorough practical investigation whether dehorning cattle was a proper process and one to be recommended to the farmers or not. The State Board of Agriculture at the session of 1887 occupied the back seats in the auditorium, and they were slow to even introduce the writer, and they were far from recommending the cause of dehorning; but they were willing to investigate, and that is just what the State Board of Illinois would not do. After investigation and after a thorough trial at their state convention this year in February, 1888, at Madison, they introduced the writer and his discovery in the following resolutions and letter accompanying:

MADISON, February 14, 1888.

H. H. HAAFF, CHICAGO, ILL.—I inclose herewith a copy of the resolutions passed in the agricultural convention just closed.

The practice of dehorning cattle is spreading very rapidly through this state, and, when properly done, no one objects to it who is familiar with its effects. I have had excellent opportunities for learning how our people are pleased with the practice, and assure you that if the present idea prevails there will scarcely be a horn left in this state one year hence.

Your lecture at our convention one year ago started the work in good earnest, and who shall attempt to stop it? I hope you will continue to study the subject and give us in an available form the results of your investigations.

Most respectfully,

W. A. HENRY.

The following resolution (referred to) was introduced by Professor Henry, and passed unanimously:

Whereas, Every year witnesses the destruction of human lives in this state by vicious bulls;

And, *Whereas*, Our horned cattle continually mutilate and injure each other, as well as other animals on the farm, with their horns, which in these days have become a useless appendage;

And, *Whereas*, Certain self-styled humanitarians have endeavored to excite prejudice against the practice:

Therefore, *Resolved*, That it is the sense of this convention, based upon the numerous reports of members present, that the practice of dehorning, as set forth by Mr. Haaff, is a humane act both for man and beast, since the pain caused by the operation is but a tithe of that which might ensue if the horns were not removed.

A vote was taken, and of the large attendance, towards a thousand farmers, sixty-six men by actual count were found who had tried dehorning; and when asked if any one of the audience had anything to say against the practice, not a single word, not a single negative voice nor a single objection was raised. Secretary Newton was the one man of the State Board of Agriculture of Wisconsin who insisted upon having the matter presented at the meeting of 1887, and he did so, fully determined that the cause should be presented even though done at his own private expense. This year, however, not a member of the Board but that greeted the writer, and was ready to applaud all that is claimed as advantageous to the farmer in dehorning.

Perhaps this side of the picture will not be complete without adding that at the recent meeting of three or four hundred farmers at Woodstock, McHenry County, Illinois, in the very center of the dairy district of Northern Illinois and Southern Wisconsin, not a farmer of all the hundreds present at the institute was found to say a word or give a vote against the practice of dehorning, either on the farm or the dairy, but nearly or quite fifty farmers were found who voted like the sixty-six at Madison, Wis. This meeting occurred in the presence of the President of the State Board of Agriculture of Illinois, and quite a number of the members of the Board, its Secretary and leading officials, and not one of them had either question or objection to present. It is a source of great personal satisfaction to the author to know that the Agricultural Boards of these two great states are substantially solid now in favor of the practice.

THE SCRIBES AND PHARISEES.

Perhaps it is not strange that so startling an innovation as dehorning should stir up opposition of the most pugnacious character and defiant kind. It is with feelings of pride, if not of gratitude, that I recount the names of those agricultural and other papers which have been veritable sheet anchors to the author in the great public contest, for nearly two years past, over the question of dehorning cattle. There is a unanimity of sentiment and a coincidence of experience from one end of these United States to the other that is simply startling when contemplated; for with little exception, not only the leading live stock and agricultural papers, and I may say without exception that the cattle-men and farmers who have practiced dehorning, or who have seen the results in the practice are the friends and firm supporters of the art as an economy on the farm, a safety to the individual and a kindness to the brute. I notice one or two opponents by name farther along.

But to return to the other side—the other picture. There is in Northern Illinois a society known as the Northern Illinois Dairyman's Association. At its recent annual meeting held at Mt. Carroll, Ill., the writer received a letter from one of his friends, and a prominent member of that association, Mr. Garfield, who read at the convention an able article on the subject of dehorning cattle, strongly recommending the practice. The writer was invited to, and did attend, that convention. The secretary of that convention expressed himself as highly pleased that "Dr." Haaff, as he put it, was going to attend. Instead of giving the question of dehorning a fair chance at discussion, as they had agreed, the chairman at that convention publicly snubbed the author; sneered at the practice of dehorning; and seemed highly elated when another member of that Dairyman's association referred in insulting language to the advent of any man on his place who should attempt to dehorn one of his brutes, and declared that he considered that it was just as proper to de-tail the animals as to dehorn them, and he would kick any man off his place who dared to talk about it to him at his home. As has always been the case, so it was at this convention. The common every-day farmers demanded to hear on the subject of dehorning, and although the author was given finally twenty minutes at an early hour in the morning, still there were several hundred intelligent farmers present to hear; and so great was the interest that the chairman at that convention felt obliged to come and say: " Please don't lay up anything as personal of the remarks that I have made." The editor of the *Dairy World*, who was present, proposed, at his own expense, to publish at once the talk on dehorning, and illustrate it by cuts something similar to those in this book. It was, however, the old story of the Scribes and Pharisees—a sort of "I am holier than thou" disposition; and it should be said to the shame of the officers of that association, that they have to this date neglected to repay the simple traveling expenses of the author to and from their convention.

The Dairyman's Convention of Wisconsin was even more seriously afflicted with an attack of the anti-dehorning rabies, but this was accounted for by one who claimed to know, in this way. Said he: "I account for it on this principle: they are all Democrats, and they don't get their eyes open until about the ninth year after any new matter is presented to them." The author will rejoice when the Dairyman's Association of Northern Illinois becomes converted to dehorning, and he will be equally glad when the Dairyman's Association of Wisconsin takes a similar position.

HORNS vs. "BUTTER POTENCY."

The following letters open up a question on the subject of dehorning unusual to the common farmer and unique of its kind. It will be understood that Mr. Hoard is the editor of a Wisconsin dairy paper, called Hoard's *Dairyman*. He is understood to be a mouth-piece and exponent of Jersey and Guernsey cattle. The G.'s and J.'s shed their horns hard. The Jersey *Bulletin* is another paper opposed to dehorning.

In order that the dairymen of Wisconsin might have an opportunity to consider this matter and discuss it if they saw fit, the author addressed the following letter to the editor of that paper. The letter is as follows:

DEAR SIR: Among the leading agricultural papers of the land yours is the only one that I now recollect worthy of notice—that is, so far as I know—silent on the subject of dehorning cattle. It is not improbable that a discussion of this question in your columns might do the public good, as has been done by other papers. If agreeable to you I will take occasion to state my views in your columns, and I will send you, if you desire, a copy of " Haaff on Dehorning." P. S. "After this letter had been sealed up, my secretary read me a line from some other paper saying that you insist that, although there may be no loss of milk in dehorning, there will be a loss of butter. Why so? How do you know? Please tell me." On the bottom of my letter, which was returned me, was the following: "Because butter is, above all other constituents of the milk affected by any disturbance of the nervous system. The nervous system finds its origin in the brain, and dehorning cannot but seriously affect both brain and nervous system. I am decidedly opposed to dehorning for male or female in any of the butter breeds, where I care for the transmission of butter potency or heredity. It may do well enough for the beef breeds. I want none of it for either Jerseys or Guernseys." To this the author again wrote in reply, objecting that the editor's dictum ought not to be taken as proof, and that his conclusions were more likely to be erroneous than not, since he had had no personal experience in the matter of dehorning cattle; and again asked this question: "Do you care to have the matter discussed in the columns of your paper?" To this second letter came the following reply: "H. H. Haaff, Chicago: Yours of the 2d at hand. If you or any other man would come forward with *facts*, not presumptions, showing that the dehorned Jersey or Guernsey bull does not lose 'butter potency' of transmission through dehorning, I will admit such facts to my columns; but I want facts, not opinions. The burden of proof rests upon you and the advocates of dehorning. We *know* what the 'potent bull' will do in the transmission before his horns are off; it is incumbent on you to show that his heifers are just as good after he is dehorned as those before he was dehorned; and the gravity of the case demands a *big showing* in this specific line, or else *the horns on my bull will stay where they are.*" These letters must be taken in connection with the editor's paper in discussing the subject. It will be observed that the editor cuts the author's legs off "up under his chin" at one fell swoop, for as dehorning is not more than two years old—in fact, is not two years old so far as the public is concerned—and as dehorning was never mentioned, much less discussed, in the editor's paper until within the last few months, it will be seen how physically impossible it would be to give "facts"—that is, such facts as are demanded, in order to prove that "butter potency" and "butter potent bulls" and "butter potency of transmission" have suffered no damage by reason of dehorning. It would be necessary, of course, to take the calves of dehorned male and female Jerseys or Guernseys and grow them to maturity, and then both as heifers and old cows show that the supply and quality of butter is no less than in the ancestors. It would also be necessary to raise the calves of these heifers or cows, and again by actual experiment on the third generation demonstrate the same fact. This could be done probably in from 10 to 15 years, by which time the chances are that the editor may be dead, and his paper forgotten. His theorizing on the subject of "butter potency" (Heaven save the mark!) is a fine specimen of special pleading. The author has seen several columns of it in his paper, but not a single word of reply has been allowed by any of the principal advocates of dehorning. But there has crept into the columns of his paper the following from one of his subscribers, which I give in full. The article is headed, "Effects of Dehorning," signed J. K. Brown, of Amy, Wisconsin, and it will be seen from this record of four cows, that No. 1 gave at five milkings after the date of dehorning nine pounds more of milk; No. 2, three pounds; No 3, one pound, and No. 4 lost one pound; and No. 4, the writer says, is "the poorest feeder in the lot." I give the following verbatim from the letter:

"Now for the effect on the habits of the animal. During the fly season the cattle while feeding would keep together like a flock of sheep to escape the flies, and we believe by that we gain many times what we may have lost at the time of dehorning. Such is the effect of one year's trial on a man whom it had taken months to persuade to allow the trial of one animal by dehorning." The editor's theorizing on the subject of "butter potency" and "butter transmission" after such an experiment will amount to but very little. He dared not omit to give the testimony of one of his own subscribers, although he was so ready and willing to summarily snub any discussion of the question by the author.

It will be seen by Mr. Hoard's letter that he expresses himself fearful that dehorning may do away in part, at least, with the pre-potency of G.'s and J.'s as butter breeds of cattle. Well, this much may be safely affirmed: by dehorning they will become "better" breeds of cattle, whether they are "butter" breeds or not. But seriously, why should Mr. Hoard and his clan fear. In his own State of Wisconsin, many men who are extensive breeders, and I believe some also importers of Jersey stock, having tried dehorning are so well pleased with the results, that they are open advocates of the process. How can the (butter potency,) as Mr. Hoard pleases to term it, be affected by dehorning? We have already shown that the circulation of blood through the membrane of the horn is very slow and of a secondary character, and that by my process of dehorning it can be so suddenly stopped that no hemorrhage worth mentioning will result. We have shown by thousands of cases, that dehorned cattle, whether they be J.'s or H.'s, G.'s or S. H.'s, care little or nothing for the operation, perform all the functions of the body immediately after being dehorned, even to rumination and copulation, and that apparently the only pain they suffer is the mere momentary shock; that it is, by the testimony of competent judges much less painful than branding, far less painful than castration or spaying, and even less so than the operation of pulling a tooth;

for cattle which are dehorned will immediately drink; they will eat, they will chew the cud, they will offer fight to others of their kind; while the brute that may have had a tooth pulled has been known to stand for hours in a dumpy condition, and the author has known cases of castration where the animal would positively faint away; and he has known like results after the simple operation of branding.

If this class of opponents propose to do away with dehorning on the ground that the butter qualities of their favorite brutes will be affected, it would seem as a matter of consistency that they should, during the last generation of personal activity, have been engaged at least in the laudable attempt to do away with branding or castration, or in fact any physical operation that may become necessary to the usefulness and betterment of the animal operated upon.

No one has ever heard any of these self-styled humanitarians complain of any operation found necessary among the breeders and cow-boys, and if the shock is to affect the "butter potency" of those brutes, why will it not equally affect the beef potency of the other breeds? It will be up-hill work and hard sledding for those editors, and their ally, the "hoss doctor," and the so-called humane society agents, to convince people that it is perfectly right (as Mr. Hoard admits it may be) to dehorn the beef-breeds and not to dehorn the butter breeds. "Butter potency" is good, but beef potency is just as good.

Gentlemen, you will find yourselves in the vocative, and it is only a question of a very short time until you will find yourselves without a corporal's guard of followers. Nor can you down dehorning by calling it a "craze."

It must be a matter of peculiar vexation to this man to hear such men as made up the Farmers' Convention of his own state declaring openly, "that in less than two years' time, there will be more cattle in the state of Wisconsin, among the dairymen, without horns than with." But the "animus of the critter" is apparent, and it reminds me of one of their number who stood up at a farmers' institute, and loudly declaimed against dehorning as inhuman and cruel, while a short time before this man had actually sold one of his neighbors a bull, and a " potent butter bull " at that, which had killed the neighbor outright a short time after his purchase. The rest of us who are unprejudiced in this matter, and who are willing to look at the subject, and treat it from the stand-point of actual benefits, know that of all the wicked breeds of cattle, the Jerseys and Guernseys cannot be matched. We know that three-quarters of all the bulls, and at certain times many of the cows, are unsafe to handle on account of their horns. We know, and so does the editor know, that every year the death record is more largely increased by reason of the "nervous" maliciousness of these cattle than from any other breeds in proportion to their numbers. We know, and so does the editor know, that when it is in proof (and the proof is abundant) that Jersey and Guernsey cows give as much milk after dehorning as before, it is simply " pop-cock " talk for any man, even though he be a " so-called editah," to reason against facts so palpably truthful in their conclusions as these. We know, and so does the editor know, that if every Jersey and Guernsey animal in these United States were dehorned there would be more and not less of " butter potent " qualities, for the reason that these breeds of cattle, which he himself claims and admits to be "of an unusually nervous character," would, by reason of being dehorned, have less fear of each other, and would hence become less nervous in their temperament, and being less so would therefore be more quiet, and hence would give more milk, and certainly better milk, and therefore more butter. All this the editor knows. All this the Wisconsin Dairyman's Association knows. All this the Illinois Northern Dairyman's Association knows, and its pesky officials know; but they are simply bound into nasty little rings, and being in the ring they propose to continue their little ringlets after their own order and dispensation of affairs; and we know, and so do they know, that within the past three months they or some of them have allied themselves to the

Humane Society (Heaven save the mark again!) so-called, of Milwaukee, the officials of which society have been making divers and sundry threats of prosecution, and one of whose veterinary surgeons, a man who spells his I's with a " Hi," and pronounces his horns without a *h*, has threatened the State Board of Agriculture of Wisconsin with prosecution for libel for daring to print similar comments made by the author in his address on this subject before the State Board in 1887. They constitute among themselves, all of them, a dirty little ring, and they are to Wisconsin and Illinois and to their grand thousands of farmers what a bevy of tree-toads are to a boom of artillery; but they constitute the only known public opposition in those states to the practice of dehorning, and they will meet the same ignominious death and burial by the public that overtook the Humane Society of Illinois something like two years since.

> If a Jersey old cus-
> Tomer stirs up a muss
>   About the " horns " of his " butter potent bull,"
> What plan shall be taken
> To prove him mistaken,
>   That others may see he's a fool?
>
> We'll stand him (agree?)
> In a wide open sea
>   Of tubs and butter crocks full;
> Then let him decide?
> When each sample he's tried,
>   Which came from his " horn potent bull."
>
> In a million of years,
> When he's shed enough tears,
>   And dried up his nonsensical sputter,
> Old Nick may slip the noose
> And let him go loose
>   If he'll shut on " bull potent butter."

The author stood one day, some months since, at the stock yards and saw a car load of bulls unloaded into the yard by themselves; there were probably no less than fifteen or eighteen big, heavy, savage-looking short-horn grade bulls. There came out of the car one little pusillanimous, black-pointed, devil-eyed, keen-scented Jersey bull. The rest of the animals, worn by their long confinement, were well content to lie down or stand in a condition of repose; they were satisfied, when food and water was given them to eat, to let the others eat and drink—taking, of course, the order of precedence, common among all horned animals, that the bosses should eat and drink first; but the little Jersey, with his black points and black-pointed horns, instead of spending his time with the food and water or in seeking repose, actually whipped every bull in that yard—nor did he rest content until he had gored and mangled in a shocking manner many of those animals, and two of them to such an extent that they actually lay down and simply groaned in misery, receiving the Jersey's thrusts without attempting to escape or offer the slightest resistance. I never saw such a case of cruelty to animals in my life. I have had the cold horns of a Jersey—one of those "nice, gentle little heifers," you know—run up my own back between skin and clothes, and I have felt myself lifted off my feet, when I escaped as by a miracle; but I confess the horror was not so intense, nor was the perspiration more plainly perceptible on my face, than in seeing this wicked little devil do his desire upon his associates in that pen, and I confess that the wickedness of heart and baseness of desire is only equalled in the case of the human brutes to whom the bulls were shipped, and who stood by and coolly witnessed "the fun" without attempting to relieve the situation or possibly in the case of the cranky editor or hypocritical "V.S.," who know that Jerseys and Guernseys, as a rule, are given to such displays of their animal ferocity when occasion offers, not only upon themselves and their kind but upon humans also.

The following letter from a Mr. Gardner, a leading official of the Orange county (New York) Farmers' Club, and a man so high in the estimation of the *Rural New Yorker* and the *Orange County Farmer* that both these great papers sent editorial representatives to witness Mr. Gardner's experiments. will be read with interest, especially because Orange county, New York, is the very center of the dairy and milk interest of that state. The letter is dated February 20, 1888, at Orange county, New York. It reads as follows:

> DEAR PROFESSOR HAAFF —There is such an interest in this section to see what you, the great original dehorner, look like, that we kindly ask you to send us your picture. We want it at our club. Well, the dehorning leaven begins to work. I dehorned a Holstein bull twenty miles away, and now two men want me to go twelve miles away to dehorn two bulls, one a registered Holstein. A singular bull accident happened here a few days ago. A party of men were riding along the highway; one of the men had a red tippet on, flying in the wind; a Jersey bull, then being in his owner's yard, saw this tippet and took it for a banter, and leaped the fence and pitched into the sleigh when they got opposite, and he then and there threw the driver over the fence, breaking three ribs for him; then the horses ran away, and the bull took after them full gallop. I think these men will be converts to dehorning Congress ought to recognize your services in the interests of humanity, and vote you a gold medal as big as a dinner-plate.
> Sincerely yours,
> M. H. C. GARDNER.

I wonder if that Jersey bull has preserved his "butter potent" qualities; or whether the "butter potent" qualities left him during the chase and settled in his horns. One thing is certain, if he lacks the power of transmitting his qualities to his progeny, humanity will be better off, even though the editor of *Hoard's Dairyman* and men of his caliber shall be displeased at the result.

The following editorial on the subject of dehorning from the *Massachusetts Ploughman* of January 7, 1888, is good reading at this point:

"DEHORNING."

Massachusetts Ploughman, Jan. 7, 1888.

It is universally conceded now that it is an advantage, both to man and beast, to have cattle devoid of horns. In the wild, natural state the horn was necessary as a means of defence. But that time has passed; and now the owner takes such care of his stock that they have no need to defend themselves, and the horn is a nuisance, an incumbrance. The amount of injury done by the horn is not easily estimated. The most serious injury is, of course, that to human life and limb. This is estimated mildly at 200 each year in the United States alone. Ugly cows and vicious bulls tear and permanently injure many whose lives are retained. Many animals are injured and a great number killed by horn thrusts. Most of the abortion is caused by ugly or accidental blows from the horns. Much of the loss in shipping cattle comes from the same cause. Thus it is readily seen that it will be a great advantage if the horns can be dispensed with. The better and more humane way is to breed off, and adopt such breeds as the "Polled Angus," "Norfolk" and "Galloway." But, while these breeds have no horns now, it is well-known that most of the polled herds were horned cattle less than a hundred years ago. (Every word true; they all had horns.)

We are attached to our "Jerseys," our "Holsteins," "Ayrshires," "Guernseys," and other breeds, and dislike to forego them for no-horn breeds; and then comes another solution to the problem, that of "dehorning."

Horses are happy with bare feet when in a wild state, but when they are domesticated it becomes necessary to give them shoes and provide for the wearing away of the hoof. The domesticated animal must have different treatment from that which is congenial to wild animals. If we take cattle under our protection it is our right and our business to protect ourselves and them. If the horns are a hindrance, which it is proven, they should be removed if the process can be done safely. H. H. Haaff, the apostle of "dehorning" reached the decision that "horns must go," after exciting experiences, which included the goring of a neighbor, his own wife, the hired girl, many colts, hogs and cattle, and even attacks upon himself. He grew to hate horns, and, finally driven to desperation, he resolved to try the experiment. He selected his best cow and sawed off her horns. It was a painful operation, resulting in much loss of blood, and a very sore head, but the cow gave as much milk as ever and lived the "best cow" for many years. His second experiment was to saw the horns off half way down; but this was awful, and he decided to kill one animal at least and find out what there was in it. Consequently he sawed the horns off close to the head, choosing of course his ugliest animal; and the result was not death. But the moment the horns were off and the animal set free, he tossed his head and went to eating. From that day he has been a "dehorner." He has received many severe attacks from opponents, but he seems to have the best of it.

Ugly animals that have been "men-killers," divested of their horns, become mild and docile as a lamb. He keeps 250 cattle, in a shed 30 x 160 feet, which would not be possible to contain 100 head if the horns were retained. He keeps them in this way warmer and more quiet and thus saves one-fourth of his hay; and as the cattle can then be kept in the barns, manure is housed and saved from the waste that is common around sheds and stables. He answers the shipping problem by asking for a car with no protuberances and no horns on the cattle. He says: "Give me the losses by horns in the states alone, on the small farms, and at the houses of every-day people, and I will become a Crœsus in wealth in ten years."

The subject seems to resolve itself into this: "Horns must go," as far as possible by breeding off, otherwise by dehorning. The dehorning need not be continued, for after a few generations the horns will cease of themselves to be. The dehorning should not be attempted by the ordinary farmer, but only by the skilled veterinarian. As far as possible the work should be done with young calves; but if grown to maturity, it may still be safely undertaken by the skilled and practiced hand; and "Haaff's Practical Dehorner" will make every man his own dehorner.

SOME LETTERS.

Possibly it will not be uninteresting to my readers if I devote a few pages to extracts from letters received from various parties all over these United States on the subject of dehorning cattle. I shall give my readers the dates of the letters, the names and addresses of the parties, so that any one can verify the truthfulness of the extracts which I offer. The following is from the editor of the *Live Stock Indicator*:

When the subject of dehorning cattle was first brought into public prominence by the arrest of its originator, at the instance of the Humane Society in Illinois, upon the charge of "cruelty to animals," hundreds of stockmen who had never before dreamed of such a method of preventing injuries among their herds, became eager to learn the process by which their unruly horned animals could be controlled, and it is a safe assertion that the prosecution of Mr. Haaff was a better means than he himself could have thought of to bring forward the advantages to the farmer resulting from the dehorning of their herds.

As early as September of that year the *Live Stock Indicator* editorially indorsed the dehorning idea, and a little later raised its voice emphatically in favor of the removal of the horns—the first journal having the courage to do so; and to-day, less than two years since Mr. Haaff's prosecution, which resulted, as is well known, in his acquittal and complete vindication, it is a fact that the horns have been removed from at least 200,000 cattle in the West alone, and that as many more will be dehorned this year does not admit of doubt. Wherever the experiment has been tried, new converts have been found, and in not a single instance have we heard of any losses resulting from the operation.

The only argument against the practice comes from that class of persons who can see only that it is a cruel operation. It is, indeed, painful, and hence they say it is contrary to Christianity as well as humanity. Yet these same persons never flinch when an animal is to be spayed or castrated, or branded with a red-hot iron, because, forsooth, that has been practiced so long that it acts upon these people about as Pope says of sin in his "Essay on Man:"—

> Vice is a monster of such frightful mien,
> That to be hated needs but to be seen;
> Yet seen too oft, familiar with its face,
> We first endure, then pity, then embrace.

But when we consider that by this one act of severity (if it may be so admitted), we prevent hundreds of other acts of cruelty perpetrated by the animals themselves, the so-called heartlessness of him who undertakes to dehorn his animals entirely disappears. Men of superior intellect are not slow to adopt new measures in controlling the brute creation (as the Almighty intended they should be controlled), any more than they are to follow in the wake of the "cranks" who first put to use steam and electricity. The world moves, and notwithstanding most great men who have brought forward new ideas, have been persecuted by their fellows, yet the very persecutions they had to undergo were the best means of carrying their ideas forward, until finally they were adopted by the intelligence of the civilized world.

So, too, will it be with dehorning cattle, and while we do not expect that the present generation of breeders of pure-bred horned animals will have the courage to undertake the work, time will eventually prove that it is a wise procedure, and though it may take long years to completely eradicate the horns, yet we are convinced that the "horns must go!"   Yours truly,

P. D. ETUE.

Mr. A. A. Carter, of Dell Rapids, Dak., says under date of Feb. 23, 1888 : "Am having splendid success. Took off the horns from a yoke of oxen yesterday. They measured 3⅞ inches at the base. Hurrah for Haaff!"

Mr. J. L. Sawyer, Gurney, Ill., says : "Dehorning works like a charm. It adds a good many dollars to the value of an animal to dehorn it."

Mr. H. C. Constance, New Richmond, Wis., says : "Dehorned quite a number of cattle this winter. Have 200 head to dehorn this spring. Have taken the hook out of several bulls—1,800 pound fellows. Have some more of the same sort on hand yet to dehorn. It is a clean thing, and you ought to be pensioned for finding it out."

Mr. E. R. Morris, Marengo, Ill., uses a couple of pulleys, and sends cut of his plan by which he makes the binding of the head an easy operation ; but he says : "If you get out a new book I want one. I got your old book of you at the State Fair at Chicago. I hope my idea may do you some good."

Mr. J. Bishop, Jr., of Delphos, Kas., writing for tools, says : "I have dehorned with your tools over 1,000 head this winter without accident. Everyone well pleased. More jobs ahead. Am making model of my portable stanchion; will send by mail as soon as I get time. Capt. Pierce, of Junction City, has been trying dehorning by Mr. B. on several hundred head with much satisfaction. It adds vastly to shed room for cattle; promotes comfort in cold weather. Dehorned cattle are friendly, crowd together, and warm each other, instead of that eternal goring, horning in large herds. It takes about a minute to dehorn an ox, but little pain, and they at once very contentedly go to feeding. Capt. Bishop is now engaged east of Minneapolis, where he has several herds to dehorn. Dehorned already in this vicinity over 300. The whole country seems interested in this late move of depriving cattle of their destructive instruments. Have talked with many persons who have had their cattle dehorned. All declare it is a paying investment for stockmen." Mr. B. has, since writing the above, sent a model of his plan, and, I regret, too late to show it here.

Mr. John Erb, of Keokuk, Iowa, says: "I am pleased to know you have had tools made for this business of dehorning; for I think you are the man to know what is necessary for success, and I know you are not guessing at what you write, but you give us facts. I have proved by experience what you say about calves, for I tried to dehorn some with a knife, and though I was very careful to get all the horn out, I found in nearly every case that the embryo horn, as you call it, was bound to come, and a second operation was necessary with your tools to make a good mulley. I have a Jersey cow fourteen years old, heavy in calf, a "devil to hook;" do you think there is any danger in dehorning her in the present condition?" Answer.—There is no danger in dehorning her one hour before labor pains begin, provided my chute is used, and care is exercised, and the hind bar which goes under the brute forward of the udder is omitted. Put the front bar under her and omit the hind bar.

Mr. E. Hermance, of Eldorado, Mo., says: "A neighbor and I have your tools for dehorning, but we are a little unsettled as to the best time to have the work done. Part of our cattle are running to straw stacks, and will be most of the winter. To do the work now (first of January), will they not get straw and chaff in their heads, as the removing of the horn exposes hollows which are liable to fill with chaff or dirt? Will it not injure them? How long will it take young two-year-olds and cows to heal?" Answer.—If these cattle have the run of good sheds in connection with their straw stacks, they will heal in from two to four weeks; and if in proper condition, and their heads are not bruised, there will be little or no suppuration of the parts. There is no danger of the orifices or frontal sinuses filling with straw, chaff or other dirt. Properly dehorned, nature will insure the healing up of the bone, as explained before.

The editor of *Western Resources*, published at Lincoln, Neb., under date of January 12, says: "There is no doubt but that dehorning is a popular move, and the time is not far distant when all grades will be shorn of their implements of warfare. We have talked with many on the subject, and have yet to see the first who is dissatisfied, or hear of the first unsatisfactory experiment, even when done in some cases by inexperienced men. Success to your new boom!"

Mr. R. H. Stevenson, of Marion, Ohio, under date of January 16, says: "I wish to ask you a question. I have dehorned nearly 400 head of cattle. I find that some of last spring's calves have horns growing again. I will ask, Do they not have to be sawed deep enough to make a hole in the skull, to make a success of the work? Answer.— No. Use the outcutter first, as explained in the directions for dehorning calves, then use the gouge made sharp on the grindstone. The outcutter will cut clear through the hide and into the skull bone. The gouge will remove the hide, horn, underlying membrane, cartilage and all, and scrape the skull, leaving the hole the size of a twenty-five cent piece, which will heal over so as to leave practically no scar. I will give a dollar per horn for each horn that you have if you will properly perform this operation as I direct.

Mr. James H. Cox, of Sandwich, Ill., says: "I am somewhat perplexed regarding 20 cattle I dehorned—20 milch cows. I did what I call a neat job. They eat well, and appear to do all right, but there is a jelly-like substance coming from the opening left after removing the horn. It is somewhat reddish in hue, and is sometimes very clear. Is it of any consequence? I have dehorned 133 head since I got your tools, and have more as soon as I can get around to it; but first I am anxious to hear from you." Answer.—If these cows were not bruised in the operation, and are not turned into the cold to stand in an exposed place part of the time, but have free access to a warm shed at all times, there will be no trouble? If they are kept in stanchions, they should not be put into the stanchions for a week or two, while their heads are healing. If there are any exceptions to this rule, it will be in the case of those whose heads are bruised, or where the blood of the animal is out of order; and in such case, the very best thing that can happen to the animal is to have a running sore for a while to clean the system out.

Mr. George Tebow, of Delavan, Kan., January 31, says: "I wish advise on dehorning cows heavy with calf. Does it hurt the cows near calving time? Is it best to leave the belly bar out so they can set down, or shall we use it? You will oblige me by answering." Answer.—Do not use the belly bar. Handle the animals gently and carefully. Have everything all ready before beginning the operation of dehorning them. Turn the animals loose at once. Watch them a little if they are quite near to calving time; you will have no trouble.

A neighbor of this gentleman, whose name I will omit

lest it should compromise him in his own neighborhood, says: "I am ashamed to say that I began dehorning with an improper saw. Some of my cattle bled fearfully, and others were terribly sore for weeks; but I have learned by experience that I can use Haaff's saw without any such liability. I know a greenhorn not far from here who had been using a common tenon saw with stiff back. He actually bled one animal to death, and another till it staggered. I taught him and a good many more the use of your saw. Let us unfurl the dehorning banner, and keep it floating in the breeze, and inscribe on it 'The Horns Must Go!'"

Mr. William Young, of Palmyra, Neb., under date of July, 1887, sends a draft of a chute. He says he finds it decidedly practical. I think it unnecessary to give it, as I believe my own fills the bill. This gentleman, like the rest, has had great success in dehorning.

Mr. J. F. Luce, of Ross, Iowa, sends description of a portable chute which I should be glad to give, as he, like the last gentleman, is a successful dehorner, and has taught many how to perform the operation. These men have dehorned their thousands.

Secretary Newton, of the Wisconsin State Agricultural Society, under date of January 4, says: "Wisconsin numbers her dehorned cattle by tens of thousands. We want you to attend our State Convention. It will receive you with open arms."

Mr. J. R. Gillus, Mt. Pleasant, Iowa, says: "I am entirely satisfied with the results of dehorning. I will never winter any more horns. My cattle are as docile as sheep— as easily handled and as easily fed. The old cows and yearlings drink freely together. Dehorning is a great reform. Now, if we can send men to Congress and to our Legislatures who will reduce taxation, wipe out monopolies and trusts, I don't see why farmers may not be more prosperous than for the last twenty years."

The editors of the *Live Stock Indicator*, under date of January 3, say: "We believe that an illustrated book by you on this subject is needed, and would make many converts. Give illustrations showing cattle before and after dehorning. This would convince many by comparison. The world, you know, is composed of cranks (else how could it turn on its axis?) but there are a great many men who refuse to own the corn, and it will take many years (just as it did Morse to make the other cranks believe in the telegraph), to convince some people that the horns are a useless appendage, to say nothing of their dangerous character."

Mr. B. C. Stoops, of Ipava, Ill., writes, asking for tools and instructions, and says: "My neighbors are dehorning a good many cattle, but they are going at it in a very reckless manner. Instead of securing the head as you do, they pull it down to the ground with a long lever, and they gets lots of blood in the operation. I am very much taken with your plan of operating." Remarks.—Any man who will hold his brutes secure to the ground by the nose or by a lever deserves to suffer, not only unnecessary loss of blood, but the loss of some of his cattle. It is astonishing how some men will voluntarily lose many dollars in the futile effort to save one.

Mr. W. L. Weber, of East Saginaw, Mich., says: "We are about beginning dehorning in this neighborhood. I have tried it on an ugly bull. It makes him as quiet as a calf."

Mr. John Ashley, of Green Rock, Neb., says: "I have read your little book, 'Haaff on Dehorning,' and also what you have said in the *Western Rural*. You have converted us to the why of it, and now we write you wanting to know how it shall be done. I think, like lots of others, the sooner the horns are taken off the better. I have lost several head of cattle by being hooked to death, and I am getting tired of that kind of work. Send me your tools."

Mr. T. E. Davis, of Eagle Lake, Minn., says: "I want your new way of securing the animal. I have been sawing off horns with a stiff back saw, but don't like it. I have seen and talked with other men who are dehorning in the same way, and they are sick of it. I have told them that they are wrong, and that I was wrong, and we have all determined to get your tools. Tell me how to stop bleeding if we should have a case—how to secure the animal without the use of a stanchion." All of which has been previously answered.

Mr. A. P. Frisbie, of Stewartsville, Mo., says: "I can do the dehorning with your tools with a good heart. I have been knocked down twice by a bull, and would have been badly hurt or perhaps killed but for timely help."

Mr. B. F. Walton, of Yuba City, Col., says: "I am very much interested in this subject. As soon as I can get your improved tools I shall dehorn 150 head of cows and 60 calves. I tried a few of the latter with a knife, but I need your tools as soon as I can possibly get them. I want explicit directions; and, as there is no one near here who has witnessed the operation, I shall be obliged to rely upon your instructions and my own judgment. If I have not inclosed money enough, send along, and I will remit promptly. I would like your cattle tags also."

Mr. E. H. Frye, of Starkville, Miss., says: "I have been reading about your method of dehorning cattle. I am interested, as I am in the dairy business. I find horns a nuisance. I have recently had several of my cows lose their calves, and I can give no other reason only that the other cows hook them. I approve of your plan. I want to know all about it. It is a new thing in this part of Mississippi."

Mr. E. D. Lindley, of Winnebago, Ill., says: "I am very much amused in reading your book at the way you came to dehorn your cattle, and the way your neighbors looked at you; but I am undergoing the same persecution myself. Three weeks ago to-day I had a valuable mare gored to death. She died in five minutes. I had refused $175 for her. Less than a week since the rest of my cattle got at and gored my best cow till she could not walk. That settled it—I am done with horns. I immediately sawed off my cows' horns. They never shrunk a drop in milk; but I did not get a good job, and as I never saw an animal dehorned in my life, I want your tools and full instructions, and will you send your bill. I cannot go to town now, or anywhere, but I am tongue-lashed on every side. Some say I ought to be mobbed; some that my cattle ought to die; some that I ought to be run out of town, and so on. But I am beginning to win. There have been two more horses killed within a few miles of me since mine. Three weeks ago this morning I was as strongly opposed to dehorning as any man in the state, but I am thoroughly converted now. Hoping to hear from you soon, I remain your most humble servant."

DEHORNER HAAFF'S REPLY TO COL. DAVIDSON.

EDITOR OF THE "TIMES": The letter in your last issue on "Dehorning Cattle," from Carthage, Ill., by Col. Davidson, is just such a letter as any gentleman might write who knows nothing of the subject he treats by actual experience. The colonel tells us that "the most diligent inquiry among farmers and stockmen has failed to elicit any satisfactory information concerning the reason for such a barbarous and unnecessary proceeding." Strong language that. Is it warranted by the facts? Let us see:

First, the gentleman admits that "he is not a farmer nor does he raise stock," and yet he assumes to call the thousand men who do raise stock, who are farmers and who have practised dehorning during the two years last past, guilty of a "barbarous and unnecessary proceeding." He might with some show of modesty, at least, have made a fair inquiry through your columns before thus openly denouncing a thing he knows nothing about save by hearsay. He does not even pretend to give us one name of any one who has tried dehorning, and who is, after trial, opposed to it. Before proceeding with the proof in this argument, I wish to notice the authority and the only one produced.

He tells us that Mr. John Fletcher, who, he says, is a senator (and I do not doubt a very honorable man. So are they all—" all honorable men ") that this learned and distinguished senator " can see no benefit to be derived," and he regards it as a " most cruel practice," and " a foolish one, too," and more, " a torture to the dumb brutes," and more still, he, the senator, " don't think that the practice will tend to produce a race of hornless cattle," and again " he has seen them bleed after they had broken off their horns close to the skull." Ah, so have we all; all of us seen them bleed on such occasions, and the senator is no better and no worse than we plain farmers in that case; I suppose both the senator and the colonel will agree with the writer that in these cases it would have been well had the animals had their horns " sawed off," or rather had they been properly dehorned (for I will not admit that the operations are identical) prior to knocking off their horns. Well now, Mr. Editor, up in Nebraska a few weeks since, I had a lot of about fifty steers out of several hundred with bad horns, perhaps most of them cattle that had been so injured by themselves or their fellows, and so in that case you see dehorning had quite a signal justification, and should not be denounced in such case as " barbarous."

2d. How do the senator or the colonel know that the practice is both "foolish and a torture," if they have never tried it. I have tried it, and I know better. One of my disciples, Mr. Evans, of El Paso, who has dehorned 2,300 head since October, 1887, says it is not " foolish or a torture." So say all the farmers he has operated for. So says Hon.

Funk, of Bloomington. So says Hon. Whiting, of Peoria. Ah, and he is bigger than a senator, he is a congressman. Go to, gentlemen. We have you there on the "Hon." side of the argument. So, too, says Mr. Webster, of Marysville, Kas., who has dehorned over 10,000 head this year; so say his patrons; so say ten men whose names I can give who have each dehorned their thousands; so says every man of their patrons; so says Mr. Richards, of Cresco, Iowa, who has dehorned over 40,000 head in same time; so says the *Western Rural;* so say the *Kansas City Live Stock Indicator, Farmers' Review, Iowa Homestead, Farm, Stock and Home,* of Minnesota, *The Ohio Farmer,* and a score of other papers, such as the *Iowa Register, Philadelphia Press, Cleveland Plain Dealer,* etc., and yet you, the senator and colonel, say us nay. Sorry gentlemen, very sorry, but you must be tolerable lonesome about now. But yet, how do these men know that dehorning calves will not tend to produce a breed of polled cattle. Fletcher (now we will leave the "Hon." out) says it won't. Haaff says it will. Fletcher has had no experience. Haaff has had experience. Haaff produces some authority; Fletcher none, save his dictum, and the fact that he is, or was once, a senator, and that don't weigh in this back-hold wrestle. This is a clear case of "catch-as-catch-can" a sort of every-tub-stands-on-its-own bottom contest. And now, Mr. Editor, as to the allegation that "Ten thousand farmers demand dehorning." Here is the proof. 1st. I produce the names and letters of half the number. 2d. The state of Wisconsin, through her Agricultural Board, has put up about 70 institutes this fall and winter, and she is teaching the art to every farmer that attends. 3d. At least 100 county papers that we know of have advocated it within a month; and, 4th, we know that nearly 10,000 copies of "Haaff on Dehorning Cattle" have gone out to as many homes among cattle men. I verily believe the number of men is twice ten thousand, and the cattle dehorned a round half million; and now, sir, I am loath to stop until I answer the other point, namely, the advantages: In August last I kept a tally of casualties published in the papers, (so far as I could obtain them), and there were forty-three, and nearly all fatal, and all but one were human, and that one was a $2,000 stallion. I might add that dehorning will save one-fourth hay to stock cattle in winter, one-tenth grain to feeders; prevent nearly all cases of abortion, save half the shed room, etc.; but the colonel and senator had better each send me their address to box 193, Chicago, and let me send them a copy of " Every Man His Own Dehorner," and they will not soon again expose their ignorance on this great subject.

Dehorning has come to stay. The "horns must go," and they are going very fast.

H. H. HAAFF.

To the Editor of " The Livestock Indicator."

Sir:—Your very readable paper has stood so firmly on the skirmish line in its advocacy of dehorning cattle as a great economic measure, that it is perhaps a questionable question whether or not I ought to be allowed to criticise any of the utterances of your paper, but I think you do me unintentional injustice in your editorial comments on Mr. Webster's letter, in which you proceed to say in substance that cattle had been dehorned a hundred years or so before my time, and that I am entitled to credit for just what I have done; leaving the inference to be plainly drawn that I have simply renewed a measure which was known and practiced long before my time. I wish to correct this error and to protest against any partnership with the Irish or English practice of "sawing off horns." During the four days' trial which I underwent against the Humane Society, it was abundantly proven that the custom in Great Britain of "sawing off horns" had been abolished by judicial, if not by legislative enactment, on account of its cruelty. We should always keep in mind the definition of cruelty. Cruelty is "the infliction of unnecessary pain." It was proved that by the English method the cattle were made to suffer unnecessarily; their horns were removed without reference to having stub horns, for no one down to my time pretended to have discovered a way of removing horns so that no stubs would follow. I wish to impress this point upon the editor and upon your readers. My discovery consisted in this; I found a place at which the horn could be removed and no stub horns would grow thereafter, and no bleeding or hemorrhage would follow of any consequence. I also discovered that beyond that point stub horns would surely grow every time, while cutting below that point would leave the orifice in the frontal bone in such a condition that it would never close. I think I should have credit for this much as my discovery, until some one can come and substantiate by proper evidence that these matters were known before my time, in which event I will cheerfully subside. I also claim that the injury to the animal when the horn is knocked off, is not so much to the wounded horn itself, as it is a congestion of the brain at the suture, by reason of springing or tending to separate the two halves of the head by a blow on the horn, the effect of the concussion; and I claim to have also discovered that the circulation of the blood in the horn is secondary, that it is consequently much less than in the other parts of the body, and I also discovered that by reason of this fact all cattle will freeze at the horns first. I also claim to have settled the vexed and much mooted question of "Hollow Horn," proving beyond a doubt that veterinary books are right in claiming that there is no such disease as "Hollow Horn," and that the old farmer himself was also right in boring through the horn at this place, because by so doing he gave vent to an ulcerated condition of the horn membrane or periosteum.

Now a word or two to those numerous persons who report their success in dehorning cattle. I am sorry that any hardware "shelf men" can be found, who will impose upon their customers a saw which they claim to be "like, or as good as, Mr. Haaff's saw for dehorning cattle." I denounce all such men unqualifiedly as outrageous frauds in making such statements. You have said yourself editorially substantially this: "A man is foolish who for the simple sum of one dollar will refuse to use a tool that is known to be right and take his chances on something that is questionable." That is enough to say and I will not occupy more space with the argument. The gentleman who reports his herd as dehorned and appearing for some days droopy, etc., was imposed upon in an outrageous manner, and it is these kind of mountebank operators who bring dehorning cattle into disrepute." It is my purpose in issuing "Haaff's Every Man His Own Dehorner," to do just what I say, namely: to make every man, every farmer, every one who desires to be, a practical first-class dehorner. I will give others' plans and cuts a full description, and my own plan alongside each other, and leave the readers of my book to judge for themselves. My mode places the animal in a position where it will not struggle; it secures the neck and head in a straight horizontal frame, without bending or twisting, so that the operator stands squarely in front of the animal—which position is, I claim, the best position for operating; and I believe that every farm should have just such a chute or frame for performing any operation almost on any farm animal, and it differs from any other chute in this respect, that no animal can cast itself in it, and it will be found that there is less machinery about its use than about any other method. I wanted to say a word about those fellows on the plains who sign themselves Jack Rabbit, Lone Star, Kil Gubbins or some sort of outlandish names, and who, I suppose, sport the generic title of cow-boy. Gentlemen, I have instructed a great many cow-boys in dehorning cattle. By the very method of your operation, I can plainly see that you are going to have a lot of stub horns on your hands, and the owners of your cattle will not be well pleased with the appearance of their unsightly stubs. Send $1.25 to the *Live Stock Indicator*, and get my new book, and read up; then write the *Indicator* another letter, and acknowledge that you have learned how to dehorn cattle. I see no occasion for tormenting the poor brutes. If, after having read the book and followed its directions, you don't admit that you have received many times its value, I will refund you your money.

Mr. James A. Davison, Practical Dehorner, of Belle Plain, Ill., writes about a year ago: " I got your little book on dehorning. Examined it thoroughly, and sent for your tools. Dehorned my cattle—115 head. Since then have dehorned over a thousand head for my neighbors, and all with entire satisfaction. At first my neighbors thought I ought to be sent to the penitentiary; one declared I was sinning against God; and in less than six months I took the horns off all the cattle he had, and he now says he would not have them on at any price. An old man of our town was horned to death by one of those 'quiet bulls.' That set men to thinking. They came to see my hornless cattle for miles around. They generally went home converted. Does it pay to have the horns off? I answer yes. It requires less than half the shed room; I save all my manure; the cattle old and young fare alike; there are no bosses; it takes a good deal less feed in cold weather, as they huddle together close and keep warm; I can ship more in a car; and if a steer gets down it is less work to get him up with no horns to catch under the others; cattle can go to market in better condition, for they are not all scratched up. I save all danger to my horses, sheep, hogs and cattle, and, last but not least, all danger to my children. I had a child tossed by a cow, and no cow of mine shall ever do such a thing again. When properly done the skin grows over and hairs out, so it would be difficult to tell it was not an original and natural mulley. No horns for me, and it will be but a short time when there will not be horns on any cattle. Send me your new book as soon as it is out."

Mr. C. A. Williams, of Mapleton, Minn., ordering the new book, says: "All that have dehorned their cattle about here are highly pleased with the result, but generally the work has been poorly done for want of proper instruction and suitable tools. Some have used hand saws, and some butcher saws. No one can do it properly without much study of the anatomy of cattle's heads, or some one to teach them how to do it. I think the cattle owners ought to be grateful to you for having so thoroughly studied and made plain to all the benefits of dehorning."

Mr. S. E. Peters, Eldorado, Mo., says: "Some of my neighbors talk of prosecuting me for having dehorned 113 head of my cattle. I am only afraid they will not do it. Some of mine that I dehorned on Wednesday calved on Friday the same week; and there is no danger in that respect, if careful."

Mr. G. R. Arnold, a Justice of the Peace of Evergreen Park, Cresswell, Colo., says: "I dehorned my herd of 175 head, four-year-old steers and calves. The tools work just as you said they would. My neighbors, who said I was a fool or crazy, have changed their minds, and I have dehorned a hundred head for one of them already. I am much obliged to you. Please let me know all new plans."

Mr. J. M. Welch, of Wa Keeney, Kas., says: "After getting your tools I dehorned over a hundred head of cattle, going strictly according to your directions, last spring. It has afforded me much comfort and profit. In my estimation too much cannot be said in its favor. I want your cattle tag, and any instruction you may have on dehorning."

Mr. Samuel W. Miller, Burlington, Kas., says: "I want your tools. Have had your little book for some time, but I think you should be more particular, and be plainer, and more of it, so that the ordinary farmer can understand better how to tie up the cattle. That should be a chapter of itself. I dehorned a four-year-bull about the time your kind-hearted and over-zealous neighbors were prosecuting you for so much cruelty. The easy way to convert such is to give them a bull to take care of, and if they are not converted after leading his lordship to water a few times, then they have a right to their opinions."

Messrs. B. & F. Tillotson, of Plymouth, Mich., say: "We send you last three issues of the Detroit *Courier*, and you will find articles concerning dehorning cattle marked in blue. We tried our dehorning tools, and found the gouge to work satisfactorily on calves. In a few days shall dehorn all our older cattle. We got acquainted with your method through the *Western Rural*. Many of the farmers are ready to follow your advice. Dehorning would have been a common thing in a short time but for the above articles in the Detroit *Courier*. They have scared off some of the farmers who are afraid of the law. For this reason we take the liberty to ask you to send an article to the editor of the above paper, which we have no doubt will rap the writer of this article into silence. We know that you can give facts, results, and reasons against which writing in the other direction is entirely in vain." The papers did not come, nor would the Detroit *Courier* send me copies on application by letter.

A controversy has arisen regarding the use of the word "dehorn," some insisting that "dishorn" is the proper word. The New England *Homestead*, of February 4, has the following editorial, which has been sent me by Professor Henry: "A controversy having arisen as to whether the word 'dehorn' is more correct than 'dishorn,' the *Homestead* referred the point to the editors of Webster's dictionary. They reply as follows: 'Dehorn is preferable.' We shall, therefore, stick to dehorn."

P. P. Holm, of Meriden, Iowa, says: "I received my tools on the 16th. On the day following I dehorned 75 head from 11 a.m. to 5 p.m. Your saw is the tool to do it with. I have dehorned 200 head with good success." Mr. H. refers to the fact that some of his cattle bled badly, and their heads matterated, and he asks the reason. Reply: Either they were frozen or bruised in the operation, or, more likely than all, got too warm during the excitement of first beginning. Of course, some of them may have been diseased.

Mr. S. M. Kelso, of Portland, Oregon, writes: "I did not have the courage to begin dehorning until this winter. I had a heifer that would hook anything that came in her way. I began with her, and I have dehorned everything from calves to five-year-old cows, and they don't lose a feed or shrink in milk."

Mr. J. T. Campbell, of Holder, Ill., says: "I had cattle dehorned on the 31st of January. I have lost one, and it puzzles me and all my neighbors. I could not see that it caused my cattle to lose a pound of flesh, but on the 29th of February this one commenced bleeding at the left horn, and he bled to death. He seemed all right for a month, and then bled to death. He seemed to bleed at the hole in the left horn. As for me I think dehorning is one of the best things that has struck the country. I would not have horns on my cattle for a dollar a head; they feed so much better, and don't seem to be afraid." Answer: If the truth could be known about this steer it would be found that he has somehow been bruised, either kicked by a horse or scared, or in some way affected so as to have bruised his head by a blow, which left the artery in the condition explained before, so that there was not sufficient contractile force to enable the inner coating of the artery to close. Had this animal been discovered in time, and the cavity in the horn and head filled with common flour, the animal kept quiet by himself, there would have been no trouble in my opinion in saving him.

Mr. James H. Cox, of Sandwich, Ill., says: "Your new mode of securing cattle for operating is splendid. I like your tools first-rate, and I see how foolish it is to use a common meat saw as I did at first. I have dehorned 255 head in the last few days. Send your new book as soon as it is out."

Messrs. Adams Brothers, of Telluride, Colo., say: "Your instructions have been of great service to us. There is nothing like experience. We dehorned ten head of milk cows before we heard from you, and they all lost considerable blood, all owing to our inexperience; and they lost flesh also. We know better now, and have no further disagreeable experiences."

Mr. J. French, of Renwick, Iowa, says: "I have dehorned over a thousand head; all doing well but one steer. He has been running a watery mucus from one side, and he is not thriving. Can anything be done for him, and if so, what?" Answer: Yes. If he has, on examination, a stub horn, take it off again, severe as this may seem, and cut a little into the matrix, as directed for dehorning yearlings. He will bleed some. Watch him; if the bleeding becomes serious use flour, and stop it. That will make a wound which will undoubtedly heal.

Mr. E. H. Warner, of Marathon, Iowa, says: "I run a herd of a thousand head of cattle, and I want your tag. Branding with hot irons is not liked by the farmers." The author thinks it horribly cruel, and now unnecessary.

Mr. J. R. Underwood, of Verbeck, Kas., says: "After a long time I got your dehorning tools. We like them much. Have dehorned 126 head. It takes three-quarters of the cussedness out of them. Our bull we dehorned weighs 1,800. He got fearfully ugly. Killed a horse worth $150 for us. Was a terror to all around. He is a grand shorthorn. He is as quiet now as a lamb, and runs with the calves and colts, and lets them eat at his manger, and three-quarters of the people here are enthusiastic over dehorning; while the other fourth say I ought to be put into prison or h——l; they don't care which. Dehorning is a great comfort to us. Our cattle drink together, and stand in the sheds together. Before dehorning it was distressing to a man with any feeling to watch them. I have lost one cow and three calves within six months by horns." He is like a good brother who writes me, saying:

"I don't think a man can be a good Christian and have horns on his cattle." And I agree with him that it is a tough job, for the everlasting roiling of the temper is something awful, and reminds me of a story, and a true one, too.

A good Christian brother, whose besetting sin had been profanity, was one day tussling with a calf, and at length, overcome and breathless, he gasped out, "You d——d little calf, if I hadn't the grace of God in my heart I'd break your d——d little neck." That was a moment of weakness; and horns on the farm are apt to produce daily provocation.

Mr. J. W. Harding, of McCallsburg, Iowa, says: "Send your complete outfit; and write me, does it hurt to put tar of iron or turpentine on the horn to stop the bleeding, or keep it from being sore? Answer: Yes. Keep such irritating stuff away from the cattle's head.

BATTLE CREEK, IOWA, IDA CO., March 12, 1888.

MR. HAAFF.

Dear Sir:—I have dehorned 450 head of cattle this spring, and have more engaged. I have a stanchion that I move around where I go to dehorn. Last week I dehorned 165 head for one man, mostly three-year-old steers; they all did nicely except one, which I noticed at the time of dehorning did not bleed as much as the rest, and that night he died; the man told me he went to the cattle yard the next morning and the steer was lying there dead, and looked by his appearance to have died without a struggle. What do you think caused his death? I dehorned seven head with a carpenter's saw, and did not like it. But the little saw goes through the horn like a hot knife through butter. I think it is just the plan.

Please answer and oblige,

THOMAS CRANE.

Mr. Wilson, of Keswick, Iowa, says: "Your saw gauge works well. I had some failures, but I think I was to blame for not cutting deep enough."

TAMPICO, ILL., March 13, 1888.

H. H. HAAFF, CHICAGO, ILL.

Dear Sir:—I have thought of writing to you for some months, but thought best to wait until I was fully convinced of the result of "Dehorning," and I am now well pleased with the result of my experiment. I dehorned all my cattle last October, and had the pleasure of wintering the herd this winter. I am now fully convinced that dehorning is a positive mercy and a blessing to the brute creation, and may success crown all your departures in life, as it has in dehorning. I was the first one to dehorn any cattle in Fairfield, and, of course, began on my own, and as a matter of course created quite a little comment, some saying I ought to be prosecuted, etc. However, dehorning is having quite a boom just now, and anybody in this town that knows enough to saw off a broom-stick, thinks he can dehorn cattle all right, or rather a great many, I should say. However, I have convinced quite a few that there is a scientific way of "dehorning," which any one with ordinary judgment, and with your tools and directions, with a little practice, can learn so as to dehorn cattle with accuracy. Nothing would induce me to keep cattle with horns hereafter. I have dehorned several of my neighbors' cattle, and they are all highly pleased with it. There are some siding saw dehorners trying to run down my way of dehorning with your saw, but I am glad to say they have made a sad failure, for, in order to compete, they were obliged to send for your saw and directions. There has been quite a number come to me for information, and I have a herd to dehorn in two days from date. I have not met with any ill-luck so far. I have dehorned cows and heifers in all stages of pregnancy, and bulls that have horns measuring over four inches in diameter. I dehorned twenty head for L. C. Russell, and he says he would not have the horns on if anyone would give him $100. I dehorned all of J. Woodward's, and he says if he had one thousand head he would not winter one horn; and this seems to be the verdict of all who have their cattle dehorned. I have saved 25 per cent. of the feed by it this winter, to say nothing of the shed room and convenience. Yours, etc.,

F. C. BERRY.

COMMENTS.

I wonder if the steer was injured by not being properly secured, by being bruised with that carpenter's saw, by a slip and a blow, when leaving the stanchion, by some other animal after the operation. I can only guess at it because I have not the facts, but this lesson I do learn: My chute would have prevented any such accident, and the price of that steer would have paid for a dozen chutes. I also see the necessity of care at all times. Dehorning cattle should not be like a " husking bee," or a " raising," or a " town meeting," a jolly gala day, but it should be a day of quiet, and careful labor, and painstaking all around.

BLUE ISLAND, ILL., March 13, 1888.

H. H. HAAFF, ESQ.

Dear Sir:—I will take pleasure in reporting to you the success I am having in dehorning cattle. The first job I did was thirteen head for one of my near neighbors. When finished he said, "There, I would not have those horns put back on again for the best hundred dollars that I ever saw." He also said that one of them had gored two cows within the last six months so they had died from the effects, and had spoiled another. He said, "No more horns for me." His name is N. B. Rexford, Jr., P. O. Blue Island, Ill.

The next was fifty-one Jerseys on the Elm Hollow Farm, owned by E. H. Rexford, also of Blue Island, with good satisfaction.

Next forty-two head for O. E. Atwood, also of Blue Island. He said that his herd were worth two hundred dollars more to him than before dehorning.

E. P. M. Wilson, of Worth, had a fine Durham bull that had gored two horses. He gave me three dollars for dehorning him, and would have given five times that amount if I had asked it.

If you have a few strong arguments that you can sum up to convince unbelievers, and answers to their foolish questions—of course, you have had them all asked—I would be very thankful to receive them by mail.

I find it hard to convince the Germans that it is right or profitable to dehorn. I talk the German language as well as the English, so I can explain the benefits so far as I have got them from your book.

Hoping that you will not tire of this I remain

Respectfully yours,

ETHAN H. WATTLES.

Remarks: My dear fellow, you are the "best strong argument" one could have. Keep right at it, and facts will bring the Germans and Yankees, too.

BROWNVILLE, IOWA, March 9, 1888.

MR. HAAFF.

Dear Sir:—There have probably been one thousand head of stock dehorned in this county within a year, after your method, and all speak favorably of the practice. I had twenty-three head dehorned three days ago, nine cows giving milk, and I can't see the least shrinkage in their milk. Quoting your own words, it looks as though "the horns must go." And we are indebted to you as the originator of a practice that is destined to be of great value to all who follow it. Very truly yours,

A. F. FOOTE.

Thanks, thanks, to you and to all—"the horns must go."

GENEVA JUNCTION, ILL., March 12, 1888.

PROFESSOR H. H. HAAFF.

Dear Sir:—I would like to ask a few questions in regard to dehorning. I commenced dehorning my calves or yearlings; after finishing seven head, I hesitated, thinking possibly I was not doing it right; some of them seemed to have holes nearly as large as a man's finger, and on others there was none perceptible. I commenced sawing right where the skin joins on the horn, or where the little crease is next the head. Was I right in so doing? All the information

I ever had was what I got from your speech at the Agriculture Society in Wisconsin.

Now, what I wanted to ask you was, Have I got time to dehorn some thirty head more and have them heal so that there will be no danger of flies? that is what is worrying me if those holes do not close up.

I think it is one of the great discoveries of the age if we can make a success of it. I have been crippled by those horns, and hundreds are constantly losing their lives.

Please answer at once and oblige, C. J. KULL.

My dear sir, we will surely succeed and not fail. There is no such word as "fail" or "can't" in the dehorning vocabulary. Cut those heads a quarter of an inch into the hair and hide (that is the matrix), and as to flies don't you be afraid. Keep off the turpentine and carbolic acid and use axle grease and cotton, and don't fear about those holes. Let them alone and nature will cure them.

RIDOTT, ILL., March 15, 1888.

H. H. HAAFF.

Dear Sir:—The dehorning tools that my father (Ellis Askey) ordered of you came all right. One week ago to-day we dehorned twenty-two of our cattle, and will finish the rest in a few days. As to the success it was much better than we expected. The cattle are all doing well, and we are very much pleased with it. I dehorned a bull so vicious that a stranger could not go into the yard where he was. Now anybody can go around him, and he takes no notice of them. It seems as if the devilishness goes with their horns. There are only a few cattle dehorned in our County of Stephenson. I am the first dehorner in our township. Some of the people that hooted at me when I talked of dehorning have changed their minds since they have seen our cattle dehorned, and are talking of having their own stock dehorned. A little time will fetch them to it.

There are some of my neighbors that would like me to do some work for them that have not good stanchions to hold them. I would like to ask you for some information to make a chute that can be moved from place to place.

Respectfully yours, F. M. ASKEY.

My chute will fill the bill. (See cut.) I commend to our Fat-stock Show people the letter of "Subscriber," below.

BUNKER HILL, March 19, 1888.

H. H. HAAFF, Esq.

Dear Sir:—Possibly you will recollect that I did buy your book and dehorning tools from you, and I can say that I have made good use of them and am well pleased with dehorning, and would not winter or feed another steer with horns, as I know it is at least worth three dollars more to winter a steer with horns than without. Dehorning is gaining ground, and several of my neighbors following my example, and some have sent after your book and tools. I did loan my tools to a neighbor, and he did break my saw before he got one horn off,—a thing I did expect.

Hoping to hear from you soon, I remain most respectfully yours,

J. H. BAUER.

GREENVILLE, ALA., March 20, 1888.

DR. H. H. HAAFF, P. O. Box 193, Chicago, Ill.

Dear Sir:—Following your advice—you remember we had some correspondence about two months ago—I dehorned my Jersey bull, Leon Gambetta. I myself admit that it was with great misgivings, and but for his fierce and dangerous character, I would never have undertaken the job. But now that the horns are off, and he is well without injury, but positive benefit, I would not have them back on him for one hundred dollars.

Of course, send me "Haaff's Practical Dehorner;" and on receipt of it, or before if you desire, I will remit you the price. It is a great pity that dehorning is not more general. Very truly, J. C. RICHARDSON.

The above is respectfully submitted to Jersey "Scribes and Pharisees."

HORNS ON THE WRONG END.

ABINGDON, ILL., February 15, 1886.

EDITOR DROVERS JOURNAL :—Among the many claims which Mr. Haaff set up in defence of the right to dehorn his cattle, he overlooked the most important points in the business, as about all mulley advocates do. But the evidence which he produced as to their being more contented, bunched together better, were warmer, required less feed, less damage from horns, advantage of shipping, feeding, shedding, etc., was all right, and not over-estimated.

Among the claims generally overlooked by mulley advocates, is the disturbance, by horns, of nature's process in laying on flesh.

It takes some time to start animals so as to get them in a thriving condition, and the process of nature should not be disturbed at any time. For instance, imagine a lot of cattle that have been fed up to a condition so as to take on an average of three pounds per day. Then imagine the condition that the last three pounds are in, or the less matured part of it, or the part that connects with the next three pounds, and see how easily the growth can be checked, and also the various ways to check it, by scaring, over-feeding, irregular treatment, and numerous other ways that all cattle are liable to come in contact with, when all experienced feeders know that horns double the facilities for disturbance. My experience in feeding cattle for thirty years has been, that every time that cattle are disturbed whilst in the above condition, it will cause loss, and the more the disturbance the greater the loss. The disturbance can be so great as to make the herd poor, and also its owners, whilst care will prevent all liabilities, except damage from horns.

Now, with all the above chances for damages, we have breeders who want to add two horns to each animal, and put them on in front, where they can do the most damage. I am satisfied that if I thought cattle were better with horns, I would have them on the other end, knowing the damage would be less. The argument that horns among cattle are no disadvantage, and do not interfere with thrift, is like the argument that searing a tender bud with a hot iron will not damage its growth.

Sawing the horns off will check the animal's growth for the time being, but the pay for it is continual thrift and everlasting escape from horn damage. The cruelty in removing the horns is about equal to two good out-door hookings, or one stable or fence corner gore; and I don't consider it as bad as constant dread or fear even among cattle. All breeders seem to be a little dull in some things. We have suffered ourselves to be educated to believe that the Fat-stock Show is the place to settle all matters in regard to the merits and demerits of our different kinds of cattle, when their condition at home should be considered as well.

Did the horned cattle that were slaughtered, and dressed 70 per cent. net, have their usual amount of hooking during their preparation for the show, or were they taken up and guarded, and fitted as a mulley would be out with his herd? Did the mulley Angus that cut 71.4 per cent. net, get their proportion of hooking during the preparation for the show? My judgment is they were hooked fully as much as though they had been with the herd all the time, whilst the horned ones were watched in order to compete with them. I have a short-horn cow that can hook at least 5 per cent. off of any steer during preparation for a show. It is her business. She hooked one cow to death this winter after the points of her horns had been sawed off. I consider that the greatest improvement on horned-cattle, next to the saw and crossing with Angus, is to watch them, and not let them frighten or gore others with their horns. It has been demonstrated at the block that they will cut 70 per cent. net by so doing. And, of course, it will pay to watch them or keep them in separate lots, or one in each pasture alone. No one ever heard of a horned steer that would cut 70 per cent. net unless it was one whose comrades were watched away from him, and kept from being gored by them. Still, sometimes, they will gore their keepers, which is the only danger aside from the danger of not cutting so much net as a mulley; 65 per cent. net is the very best one could expect from a natural horned production, fed loose in the field; 50 per cent. net is good in a pen, loose together; and 70 per cent. where they are carefully watched. And to feed one with such as my cow, one need not look for more than a lacerated hide in the spring.

Now, all considered, Mr. Haaft's account of saving 20 per cent. of feed by the different advantages of dehorning is not unreasonable. Now add the difference in per cent. between guarded ones, such as are to compete with mulleys, and ones fed in a hook-as-they-please lot, and you will find 20 per cent. difference in gain. And the more they are like my short-horn cow, the greater the difference, and so on until you come to the hide alone, without any per cent., as mentioned above. SUBSCRIBER.

To the Editor of the Farm, Stock and Home:

Sir—Some years ago when I was in the throes of personal prosecution and individual persecution on account of dehorning cattle, one of the farmers' papers of Illinois used the following language: "Like all reformers, Mr. Haaff has to be a martyr to the cause. Every day or so somebody starts a fresh lie on him, and it spreads around and is handled over and over till it is worn out; then they start another one." From the unsuccessful finale of all personal attempts of this character which have matured and come to grief within the last ten years, I am called upon to combat a new proposition, which, as the younger Pitt described an attack upon himself, is urged upon me by a correspondent in your columns with "so much spirit and decency." We are told that I "am making money out of dehorning." This seems to be the drift of that article, and the moving cause the writer had in view in penning the same was to expose my motive in so doing, and that he may give full effect to the sting which he seeks to produce, he prepares your readers by telling them at the outset that he don't propose to hold a controversy with me, and he also proceeds to show what a modest and amiable gentleman he is by stating that he was present at the State Convention of farmers at Madison, Wisconsin, in February last, and that when the question came up as to "indorsing Mr. Haaff's method of dehorning cattle," he, like the brave and true man he is, kept his seat, and "didn't vote." Now, I leave your readers on this line to draw their own conclusions as to the animus of this party. I have mislaid his article and shall have to reply from memory, and I am delaying for a few days the issue of my book on dehorning to answer the objections of this and other critics, and to denounce as a fraud the operations of one Wicks. This man Wicks takes another tack. The previous correspondent to whom I have referred—I think the name is Phillips—objects to my method because he says in " about six years the progeny of dehorned cattle will 'begin to give us a race of cattle subject to catarrh and diarrhœa.' " I have to assure your readers and the public through your columns against Mr. Phillips' "catarrh and diarrhœa" on the one hand, and against Mr. Wick's thievery on the other. This man Wicks is advertising himself, principally by circular, as I am informed, throughout the Northwest and West, as giving a "new method of dehorning cattle," and he sends out a circular, he says, of twenty pages for ten cents. Your readers and the public generally should be warned against him. All that is most valuable in his circular is stolen bodily outright from my copyrighted work, " Haaff on Dehorning," and as to his "new method" it is simply a reproduction of the one described by me in that book, in which I cast the animal, and run a knife around the base of the horn, severing the hide from the horn before using the saw—a method bloody and painful in the extreme to the animal, and which, coupled with casting and tying the limbs of the prostrate animal to a post from behind, is both unnecessary, cruel and injurious.

Now, one word on the "catarrh and diarrhœa" business. Mr. P.'s proposition reminds me that we have had three attacks on three various occasions from substantially the same source. This is the Milwaukee and Wisconsin Jersey and Guernsey ring. Their first objection to dehorning, made some years ago, was that it would destroy the value of the bull to dehorn him. This objection was virtually disposed of, because I produced abundant proof that bulls that had been dehorned for years by myself and others were equally and every whit as good getters as before being dehorned. Driven from this position during the last year, they have taken up another, namely: that the "butter potency" of the bull, as they choose to call it in the abundance of their wisdom, would be injured. I think in my new book I will abundantly explode that doctrine, and knowing that I will do so they have put your correspondent forward with still a third objection, and that is the "catarrh and diarrhœa" subterfuge. There is nothing to it. I have dehorned cattle on my own place, and kept them by hundreds, yes, by thousands, for eight years last past. I was speaking with my boys last evening on the subject, and the boys both agree with me in saying that neither among milk cows which were stabled, among fatters which were stabled or shedded, or among stock steers and young cattle, did we ever have in these years, either earlier or later, a case of either catarrh or chronic diarrhœa that we can now recall. We know that dehorning reduced our percentage of abortion to almost nothing, so that out of 125 calves last year, including a good many heifers, we lost only two. I don't believe that any herd of horned cattle ever showed so small percentage of loss. My boys say that the fourth crop of calves from the imported Hereford bull Dauphin since he was dehorned showed last fall at the age of six months a very marked falling off in the tendency to produce horns, and the boys say that the previous year's crop of calves from Dauphin, about seventy-five of which were dehorned as calves, did not produce a single stub horn. The value of our own experience in this line may be better understood when I say that I kept a dairy of breeding cows that were dehorned Shorthorn cows and heifers. I kept them solely for breeding purposes, so that some of the cows that were bred eight years before are some of them on hand now, and the best calves of each year's crop were kept for breeders, say from eight to twelve a year. This would leave us, as near as we can tell, from fifty to eighty that had been dehorned for eight years, and about ten we think as a selection of breeders from each year's crop. I think, Mr. Editor, it will be conceded that this experiment is sufficient to settle the diarrhœa and catarrh tendency, except as to the brains of those who are determined to object to dehorning, per se, whether right or wrong. I shall never attempt to cure their affliction. It is, in my judgment, a hopeless task.

Now, I would like to ask, since I am through with this branch of the subject —I would like to ask this combine of editors and Vs.'s and Humane Society men and such men as your correspondent who allies himself to them, what object can you have, and what earthly good can you do by pursuing such a course as you do with reference to dehorning cattle? You admit yourselves, or your leader does for you, to use his own words, "that it may be a good thing for beef cattle," but he says, " I will have none of it for Jerseys or Guernseys." I cannot see the bent of your purpose. Can't you take just as much pride in your Jerseys and Guernseys if they are deprived of their power of killing each other and destroying human life? If you are fair men; if you are what you claim to be, in its true sense, men who are humanitarian in your tendencies and aims, you ought to be the very first to desire the removal of these unnecessary auxiliaries—horns. I confess I can see but one motive. Your Humane Society expects to get to itself a big name, and hence it has, to use the fair conclusion of your correspondent, "sent over to Great Britain to find out whether dehorning is right or not." Your editors and the publishers of your papers are mad because they are not being paid by advertising or otherwise for pushing the cause of dehorning, and your Vs.'s are afflicted alike with a disease of the brain and of the pocket. By opposing dehorning they make themselves singular in the eyes of the community, a d they may expect to gain in a professional way by casting odium upon the practice. It looks to me that you have mapped out for yourselves a very rough road to travel.

H. H. HAAFF.

Such men cannot deny that dehorning is a success, nor that I gave it to the world in its present shape. Even the word was unknown in print until I used it. Show me the printed word prior to my time—when and where it was used. I created a necessity for the word. The Irish and Scotch, it is true, sawed off horns, but they did not dehorn.

Show me where it is written how to dehorn and have no stub-horn follow, or where it is printed how to cut and leave the opening into the frontal bone with a permanent hole after the wound is healed up. Who, prior to my time, could say that a stub-horn may be so cut as to give a mulley head, and yet the neglect to properly re-cut (by cutting a trifle below the first cut) is what will (if anything ever does) bring dehorning into disrepute. Stub-horns and the neglect to recut may do it, but these men can't do it.

EXTRACTS FROM "HAAFF ON DEHORNING CATTLE."

Dr. Cutts' Letter.

GENESEO, Ill., Jan. 31, 1886.

HON. H. H. HAAFF.

Dear Sir—In the suit of the State Humane Society against yourself for cruelty to animals in sawing off their horns, recently on trial in this city, the testimony of experts in behalf of the prosecution went to show that the operation was one of great cruelty, inflicting severe pain upon the animal entirely disproportionate to the benefits expected to be derived from it. If this is true, the case should have been prosecuted and an attempt made to put a stop to the practice. But the overwhelming weight of testimony on your side from farmers and others accustomed to the care of cattle, who had either seen the operation of dehorning or had performed it themselves, that the animals did not apparently suffer much pain at the time or afterward; that they manifested no symptoms of shock, but partook of food and water immediately; that in milch cows the secretion of milk was not in the slightest degree diminished or changed, goes very far to prove that the operation is not so severe as has been generally supposed, and that the testimony on the opposite side was given more as the result of preconceived opinions and theories rather than from actual study and observation.

It is quite possible that the nervous system of animals becomes less and less sensitive to pain in proportion as they descend to lower grades, but of that we can have no proof, except from the behavior of the animal itself. Assuming in this case that there is no difference in sensibility to pain, it becomes important to consider the anatomical construction of the horn, in relation to its supply of nerves and blood vessels. It may be stated as a rule, that while the nerves that are sensitive to pain are more generally distributed over the surface of the body, they are distributed with increased supply to parts that undergo rapid waste and repair, and are diminished where these processes take place more slowly. The same statement may be made with regard to the blood vessels which supply the materials for nutrition. No physiologist will assert for one moment that there is any very considerable circulation of blood within the horn, or any active process of waste and repair going on there to require such a circulation. It would be to contradict all the results of observation and experience. On the contrary, the circulation is extremely small and the changes are very slow, so that we might inferentially from these facts alone come to the conclusion that the nervous supply is very limited also. It does not change this conclusion if we assume that the function of nutrition is controlled by the sympathetic nerve, which is not a nerve of sensation; that would of itself render the supply of sensitive nerves still less necessary or important.

The horn of an animal is its weapon for attack or defence, subject to rude shocks and heavy strains and to occasional loss by violence. It would seem improbable from the very nature of its functions that it should be endowed with a great degree of nervous sensibility, which would be a disadvantage rather than an advantage to it.

No one can for a moment suppose that there is any susceptibility to pain in the outside shell of the horn. A blow upon it may, by jar or conduction, affect the more sensitive parts at its base; but it is itself as absolutely incapable of sensation as is the hair, or the free border of the nails; belonging as it does to the same epidermic structures that contain neither nerve nor blood vessels, consisting wholly of exuded or formed matter, that undergoes no further changes than to be cast off. The central part of the horn in young life consists entirely of cartilage which is separate and independent of the frontal bone. This becomes by age converted into true bone, by the same process that takes place in fœtal cartilage, viz.: the deposition in it of mineral matters, chiefly salts of lime. Neither cartilage or bone are sensitive tissues; that is, they are not supplied with sensitive nerves. They may be cut, or sawed, or gouged, both in health or disease, with very little pain, as every surgeon knows. The pain that is felt in certain cases of disease comes from the pressure of exuded matter involving nervous filaments in their periosteal coverings, rather than from the substance of the bone itself, and so also of cartilage.

There can be no part of the horn, therefore, above its base, that is sensitive or painful in the operation of sawing it off, except a thin circular layer between its outer shell and the inner bone. This layer is formed by the corium or true skin (its outer division forming the shell), united to the periosteal covering of the bone, and probably in no case exceeds one-eighth of an inch in thickness. It is the only part of the horn supplied with nerves, and it is supplied from the same source that supplies the skin and muscles of the forehead and temples, viz.: from the supra-orbital branch of the ophthalmic nerve, which itself is a division of the fifth pair, with perhaps some terminal filaments of the facial nerve. Of course the division of this tract must cause some pain, but there is no reason to suppose that it is any greater, if as great, as that caused by a section to the same extent of the same nerves in the skin of the forehead. The sensitive surface exposed in sawing off a horn three inches in diameter does not exceed one and one-eighth square inches ($3 \times 3 = 9 \times \frac{1}{8} = 1\frac{1}{8}$) of a cross-cut section, that is always less painful than where the terminal ends of nerves are left exposed or inflamed, as in branding with hot irons, to say nothing of the much greater surface involved in the latter operation.

No nerves can reach the interior of the horn from the nasal cavities, for the reason that there is no communication between the two until after the age when the frontal sinus becomes developed and the apophysis, first of cartilage and then of bone that forms the central horn, begins to atrophy to form these cavities. That button of cartilage in the calf that eventually forms the horn is entirely separate from the frontal bone, and only becomes joined to it by age. Its imperfect or arrested development gives origin to the breed of mulley cattle—cattle without horns.

Now, the arrest of development in any special organ, so general as to become a race characteristic, is regarded among naturalists as *proof* that the part so arrested was neither highly organized, important, nor even necessary to the animal. The inference is unavoidable that the part suppressed could never have been highly endowed with nerves, or it never could have been suppressed. Nature does not make mistakes of this kind.

We have said that the nerves of the horn come from the same source as that which supplies the skin of the forehead. There is no other source from which they can be derived. The sensibility of the horn, therefore, cannot be greater than the sensibility of the forehead, but, on the other hand, it may be greatly below it, owing to the diminished supply of terminal nerves. On this point *no proof* can be offered, because, so far as I know, no actual demonstration of nerves within the horn has ever been made; they are assumed to exist by analogy only.

The skin is more freely supplied with nerves than any other part of the body, and it is the part most sensitive to pain, but even this varies in different situations and is by no means uniform. A surface denuded or inflamed is painful in proportion to its extent; but if the same surface is at once covered up and protected from the air, the pain becomes comparatively slight. It is not so much the violence, the laceration, the local injury, as its exposure to the air afterwards that renders it painful; and out of our knowledge of this fact has grown up the modern method of treating burns, scalds and similar injuries, by covering them with impervious coverings.   J. B. CUTTS, M. D.

This is a capital letter, especially as the doctor "changed his mind upon investigation," after an interview with the author.

There are several questions yet in an experimental state that I had hoped to present here, and declare ready for the public. The matter of warming water for cattle is one. The matter of tags for cattle is another. Both of these questions will find a place soon in the agricultural papers, and will be solved in a manner satisfactory to the farmer and to stockmen.

In presenting a new edition of this little book to the farmers and stockmen of the United States, it seems the proper thing to give the present status of the art, the objections now urged, and the future outlook.

More than two thousand farmers have dehorned their cattle in the northwest alone during the fall and winter of '86. The demand for books, for tools, and for the personal presence of the author, is greater this April than at any previous time since dehorning began. Will the practice become universal? Well, time alone can determine that question. This much seems settled. No one who has tried it goes back on it. All seem to endorse it. No injury has followed. We have never heard of any loss by it. True, our friend, Lan Waite, of Sycamore, dehorned a Jersey bull and the bull died, but all admit that dehorning was not the cause. If men will "go it alone" when for thirty cents they can "read and know," we do not see who but themselves is to blame. Two other instances of death have come to our notice, but in both cases ignorance or carelessness was the cause. Most of our readers are like our friend Moses, of Geneseo, Ill., who says, "We dehorned our herd on our Nebraska ranch, and we would not have the horns on again for five dollars a head."

We see no reason why dehorning shall not obtain all over the land. It is certain that dehorning:

1st.—Will save 200 human lives yearly.
2d.—200,000 cattle and horses yearly.
3d.—Great numbers of hogs and sheep.
4th.—One-fourth the hay in winter.
5th.—One-tenth the corn to feeders.
6th.—One-half the shed room.
7th.—One-half the manure.
8th.—Nearly all loss of calves by abortion.
9th.—All loss in shipping cattle.
10th.—Profanity enough to sink a nation.

All this yearly—giving an aggregate saving in dollars and cents of over $10,000,000 annually in Illinois alone, and $100,000,000 in the United States. If anyone has a greater or a more modern plan of benefiting the human and the brute creation, we have failed to hear of it. It was well said by one, "As I think of it, its advantages become more and more apparent, until I am astonished that I did not think of it and put it into practice years and years ago."

Opposition.

But are there no opponents? Of course there are. There can be no good thing without. There is the "Champion Liar" of Henry county, and his ilk. Nothing escapes their carping, dirty slang. With no brains and less decency, they make up in noise what they lack in sense, and remind us of old Sam Johnson and the young man who said, "Well, I must live, you know." "Well," said old Sam, "I don't see much need of it."

Then there are the Scribes and Pharisees. They don't believe in it—"Not though one rose from the dead"—no; of course not.

> Wise in our own opinion, we,
> And wiser we don't mean to be;
> Tho' seven men can render reason,
> Their talk is heresy and treason.

Then there is that other class, who "have the gentlest bull you ever seen," and they don't believe in it, and they won't until a cold horn is run into them—when possibly it is too late. "Oh!" said one of them, "what a captious fool I was, and how it all came to me as I was lifted up and sot down on the fence. I'd 'a gin all my old boots and shoes then to have been a minute older." At the start another had the actual presence of mind to "lift himself right up off that horn, which was run into me more than six inches, and I don't want any more."

But Cattle Will Butt.

True, cattle will butt, and their butting is bad, and much to be avoided; but the absence of horns will always commend itself to the thinking person. "If the head must come, let it come minus the bayonets," said another.

Carving Calves' Horns.

No man can successfully dehorn calves with a penknife or jack-knife. Get the printed directions, and follow them; this knife business means more horns. Use the gouge as directed.

Dehorning With Shears.

One man writes that he has "improved my method. I use shears similar to pruning shears, and the horns come off in a jiff." Yes, and they will grow on again. You cannot dehorn cattle properly with shears. Use the saw, and follow the printed directions, and no stubs will result. The operation is so different in yearlings and two-year-old cattle than in older, that you must understand just what to do in each case, remembering that "any time but fly time" is a safe rule as to time, and remembering also that shears will damage the head.

Comments of the Press.

The pioneer paper in all the discussion that has resulted during the past years is the *Western Rural*, and it deserves honorable mention for the free use of its pages. It has had some hard shots along with the rest of us, and has warded off the blows with its head. It is a capital medium of communication for the farm and the home. May it ever thrive. It says of the practice: " Ninety-nine out of a hundred who dehorn cattle say that it is a good thing. The evidence presented has so impressed us in its favor that if we were feeding cattle for profit, in these times of low prices at least, we think we should take off the horns as quickly as we could get Mr. Haaff's saw." It says further:

THE CRUELTY OF DEHORNING.

We are asked by a correspondent if it is cruel to dehorn cattle? This subject has been thoroughly discussed. We have no evidence that it is as cruel when properly done as some other operations which we perform on our animals. It doubtless hurts some. But it needs to be done properly. According to Mr. Haaff's method—and his method is the only one worthy the name of method that we have yet heard of—there is very little suffering. It is amusing to see the number of people who have been dehorning cattle all their lives and whose forefathers before them dehorned, now that through Mr. Haaff and *The Rural and Stockman* the subject has been brought to public notice. There are so many who have been at the business, and for so many years, that we almost wonder that when Mr. Haaff got ready to dehorn, he could find a single horned animal, even a Merino ram, in the whole country. There may have been, and doubtless were, men who sawed off horns in Great Britain, but no one prior to Mr. Haaff was able to give a reason or point out the place and manner of operating. But Mr. Haaff was the first man to do more in the way of removing horns than was done by about such an operation as knocking off a horn with a club would be.

The latest claimant to a thirty-year experience in dehorning is a Nebraska man. He is the last man to say after the woman has killed the bear, "See what Betsy and I have done." If Mr. Haaff had dehorned cattle as this gentleman describes the method, the Illinois Humane Society would have had him dead to rights when it prosecuted him for cruelty to animals. If the method described by this gentleman were the only method, *The Rural and Stockman* would denounce it, first on the ground of cruelty, and second on the score of trouble. He gets the animal down, ties it, saws away, drawing blood in large quantities, frequently exposing the brain, and is compelled to cover the opening to protect the brain, etc. Then he introduces a sentence ridiculing Mr. Haaff for "introducing dehorning in the United States!" Mr. Haaff has not introduced such dehorning as that in the United States or elsewhere. *The Rural and Stockman* has never before given publicity to any such method. The truth is, that all that is known of scientific dehorning Mr. Haaff has taught, but we presume he will not have the slightest objection to the probably large increase of the number of dehorners to let them tell it, especially if they describe their methods.

Don't draw any head down. Don't do it. And above all things, don't put a thing into the nose to hold the animal. It is cruel and unnecessary. Lift up the head and secure as shown in cut. To which let me add this: don't bind any animal's legs in dehorning; it is cruel, and will cost you dearly if followed up.

I add the following cut from the *Rural* for the sake of giving Mr. Heath's letter:

Letter from Mr. Haaff.

Editors Rural and Stockman:—I never wish to be a pioneer again. The man at the front gets all the kicks and no pension. Here is a gentleman who writes from Maize, Kas., complaining of me that I didn't answer his questions, and informing me "that you don't know it all," etc. I have written him a personal letter asking to be told again, and that I will be sure to reply. I am sorry to offend anybody, and, I guess I do a good many times by my blunt ways; but it seems to me as if the dehorning question had been thoroughly discussed, and I do believe that every man who uses my saw and gouge, who gives them a fair trial and waits results, will say as Mr. Heath does, whose letter is appended hereto. I simply asked Mr. Heath, who by the by is a wealthy banker at Lafayette, Ind., to wait until he was convinced one way or the other and then write me. He owns over a thousand head of Hereford cattle, and they are good ones, too, and he is not afraid to put up bulls worth hundreds of dollars apiece and have them dehorned. One man's evidence like his ought to settle something anyhow. Here is his letter, and let me say it was written months after I saw him, and without any call by me at all, or a word more than said at parting last spring:

Lafayette National Bank, Lafayette, Ind., September 28, 1886.

H. H. Haaff, Esq., Atkinson, Ill.—*Dear Sir:* I saw my dehorned cattle a few days since, for the first time since last May. It gives me much pleasure to inform you that they are doing well, and that the result of the experiment has proven highly satisfactory to me. As an evidence of my gratification over the result, I shall in a few days dehorn all my present year's calves and many of my more mature cattle. I have not considered it judicious to communicate these facts to you until every doubt upon the subject had passed from my mind. Congratulations are extended to you by reason of my thorough conversion to your theory and practice of dehorning.   Respectfully yours,
John W. Heath.

I do hope your readers will not allow their prejudices to stand against pocket books. Allow me to close by adding again, dehorn your cattle, young and old. You will save one-fourth your hay; one-half your shed room; one-half your manure; all loss in shipping; all loss of calves by abortion; all loss of life to cattle, horses, sheep and hogs; all loss of human life; and, what is best of all, your own temper.

And I place after it this communication to an agricultural paper:

DEHORNING CATTLE.

MOORHEAD, MINN., February 10, 1887.

I have had experience in dehorning cattle for several years, and for those about to try it, would say I use tree-pruning shears, with which I can clip off a horn up to one and one-fourth inches thick in a moment. For stock six months old and under, I let one man hold the animal's head steady with a common rope halter, another man holds the animal close to a fence or a board partition. For larger animals we have a ring in the floor to draw the rope of the halter through, and a strong stanchion to fasten more securely. When horns are more than one and one-fourth inches long the saw must be used, and certainly all stockmen will agree with me that it is highly humane to dehorn the brutes, for after the act is performed, docility reigns supreme. No master of the trough among a lot of "sore heads." Yours truly,

F. J. SCHREIBER.

Wise Mr. Schreiber, where have you had your talent buried these "several years?"

I add the following letters from the W. R., because they are not only good, but come from men—plain, every-day farmers—who do not write often for the papers:

FOR DEHORNING.

EDITORS RURAL AND STOCKMAN:—I am a farmer and don't often send in anything for publication, but when I see a man like W. J. S. talking of something that he knows so little about, and without any experience in dehorning telling how men have tortured cattle by chopping off horns, I must say a word. Mr. Ignorant claims he has had cattle fifty years and never had any horned to death or give bloody milk, and that he kept bulls and never had one that was dangerous on his farm. He says he has a short-horn bull and is no more afraid of him than of a lamb. Now, this may be all so, but is this the case with hundreds of other men that are handling bulls and cattle? Not long ago I purchased a short-horn bull calf for $200. After three years he began to use his weapons, and was not to be trusted at all. Finally I traded him for another bull, but the man that I traded with was very unfortunate, for the bull got loose in the stable and killed a fine mare on the spot, worth $150 cash. Now, was that bull's weapon worth that? No, nor himself either. The bull I got in exchange for mine was more vicious and ugly than my former one. I would like to ask Mr. Ignorance if these bulls were lamb-like? Another ugly bull we had in the pasture would bellow and paw the ground when people were going along the road. Is that lamb-like? A few days ago a cow hooked another into the barbed fence, breaking off the wire and posts. Another ran after a colt, trying its best to catch it. Two years ago we turned some horses into the cattle yard for exercise. In a short time one fine animal, a mare, lay dead in consequence of cattle's weapons. She was worth $150 at least. How is that for lamb-like stock?

Why should this man condemn Mr. Haaff when so much good has resulted from dehorning? I consider Mr. H. has saved the country thousands of dollars every year, to say nothing of the human lives saved. To test the dehorning, we began over a year ago to saw off some of our vicious cows' horns, with very fine results. Then, afterwards, we made fast in the stanchions several head and dehorned them in a few minutes. I consider the torture to amount to nothing, not so much as to castrate an animal, for when let loose mine soon went to eating. Nor did the cows shrink in their milk. But how it quieted them down! Their ugly disposition had gone. No more killing horses or running after colts, no more goring calves; no more tearing down fences; no such hooking and pushing at the water trough as formerly. Now they do appear lamb-like. I am not in the least fear of going before them, or close to their heads. In regard to those dehorned a year ago, the horn at the place of sawing off is healed over solid, so that no bad effects can arise from it. Let me again repeat that Mr. Haaff has been a great blessing to his country in introducing dehorning. As for me, I don't want any more horns on my place. It makes no difference how lamb-like they appear. Mr. W. J. S. says God gave man dominion over the beasts of the field. That is it exactly. But could he have dominion over their weapons? Their weapons were made only for their wild state, but it belongs to man to improve on these weapons. Suppose because it hurts a little we should not castrate, but let our stock of all kinds go just as they were created, what would we have then? This dehorning is a great improvement in subduing them, as it also improves their looks. Dehorning is not going to stop. L. R. HILLMAN.

CRUELTY TO STOCK.

EDITORS RURAL AND STOCKMAN:—During the past year or more, while perusing the columns of *The Western Rural*, I have become much interested in the subject of dehorning, as advocated by H. H. Haaff, and it is but natural that when one's attention is called to this subject for the first time that the vision or remembrance should come before his mind of some cow or steer which he may have seen whose shell horn has been slipped, leaving the sensitive tissue which covers the bone horn exposed and bleeding profusely. With one class the subject, without investigation, is thus dropped and the verdict pronounced. Oh! cruelty. Now, while we would not approve of every advanced theory without investigation, it is not in accordance with the true enlightened idea of an American citizen to dimiss a subject without first giving it careful thought and try to arrive at an intelligent decision. Practical dehorning was new in this section of Iowa until adopted by the writer, under the instruction of Mr. Haaff. Yet the subject was familiar to the minds of an intelligent class of farmers, especially the readers of *The Rural and Stockman*. I find also that those who in the past have been most careful to provide for the welfare and comfort of their stock are the ones who most readily approve of dehorning. This is not strange, for when one has provided at great expense sheds for their protection, it is not pleasant to learn the fact that you have merely provided shelter for a few strong old bosses, while the weak and timid ones stand out in the storm and look in the same with the feed racks and water trough.

As an example of those who raise the cry of cruelty, I am reminded of a man of my acquaintance whose cattle had been known for a number of years by having their ears and tails frozen off from exposure. I notice several writers on the subject of dehorning advise their readers to "procure a good sharp saw and blaze away." Now, I think, such instruction is rather too brief. I would say, learn all you can conveniently from those who have given the subject thought and have also had some practical experience. Then provide suitable ropes and tools, also stanchions properly constructed, so that the work may be done in a quiet yet expeditious manner. This is especially important when dehorning milk cows or those heavy with calf, and if the rules advocated by the originator, Mr. Haaff, are practiced, the work may be done without even a shrinkage in milk, or any ill effects resulting. The writer has now taken the horns from two hundred and fifty head, about one-third of which were milk cows, and these la ter have expressed themselves so plainly through the milk pail that I am now convinced that the pain inflicted is not worth considering in comparison to the advantages resulting either from a humane or financial standpoint. I am in favor of breeding and gouging the horns from the calves, and sawing from the old cattle, and with *The Rural and Stockman* I am in favor of dehorning the devil by educating the youth, and removing the horns from the intemperate by suppressing the liquor traffic, thus saving more than one-half the misery caused to man and beast as a result of horns. B. F. R.

Dunlap, Iowa.

The following clippings from the *Farmers' Review*. Another stanch friend of dehorning presents points of interest to all:

Mr. H. H. Haaff attended the Wisconsin farmers' convention held at the state capital last week on invitation of Professor Henry and others connected with the state board. A hearing was arranged for him on Tuesday P. M. in the assembly hall, the legislature adjourning for that purpose. The hall was packed to its full capacity. Of course his theme was dehorning, which he illustrated by specimens of skulls minus horns and horns minus skulls. Professor Henry's published experience in dehorning had prepared the way for a favorable reception of the plan, and judging from the reports of the meeting in the Madison papers Mr. Haaff seems to have carried everything before him, fully four hours having been taken up by his speech and subsequent reply to questions before the meeting was willing to disperse. There can be no question but dehorning has come to stay, and at no distant day will be generally practiced, but upon young calves instead of grown animals, the wearers of horns having meanwhile all disappeared.

The *Review* further adds:

We think it folly to claim that the operation of dehorning cattle does not cause suffering. The real question to be considered is, "Are the advantages resulting from dehorning such as justify the infliction of whatever degree of suffering attends the operation?" There are other operations, the propriety of which no one questions, which also cause suffering to animals, such as castrating males, spaying females, branding, cutting off the tails of lambs, making ear marks, etc. These operations, under most conditions, are regarded as necessary, and so justifiable. In case of dehorning, if properly performed, we are satisfied the suffering is not as great as is claimed by many. The argument against dehorning because the animals were created with horns, would, if carried to its legitimate conclusion, prevent the castration of all male animals as an interference with nature; and if adopted and put in practice would in about three years bring about a condition of things among our domestic live stock which N. L. H. would not find it pleasant to contemplate. In this, as in many other things, prevention is better than cure, and it is better to operate on the four-weeks-old calf, so as to prevent the growth of horns, than to let it grow up and then have a serious tussle with it to take the horns off. The comparison instituted by our correspondent between the suffering caused by taking off the horns of an animal and cutting off the ears of a human being is not a fair one. The horn is not supplied with sensitive nerves as is the ear. The only sensitive part is the thin membrane enveloping the inner bony core, not thicker than a sheet of paper. Neither the outside horn shell nor inside bony structure are supplied with nerves, and besides the nervous system of the bovine is not as delicate and sensitive as that of the human family.

It may be said right here that the operation of dehorning is more severe on the calf than at any other age. The printed directions sent with each gouge should be carefully followed, or stubs will appear.

The *Live-Stock Indicator*, of Kansas City, comes out square-toed and flat-footed, and says " horns *must* go." Hear it:

Horns Should Go.

The *Live-Stock Indicator* is thoroughly convinced that horns on cattle are a cruel and costly nuisance, which the breeders of the future have no valid, sufficient excuse for tolerating or perpetuating.

Again it says:

Henceforth farmers and stockmen will be seeking information about dehorning calves—in what manner and at what age it is best done; and the agricultural colleges and state veterinarians should be utilized as mediums for disseminating correct knowledge on the subject. The colleges raise more or less cattle and afford opportunities for giving their students ocular demonstrations of the work, while the state veterinarians should, at farmers' institutes and similar meetings, give not only brief lectures on the subject, but perform in public the operation on calves that would be brought there for that purpose.

And again:

The Horns Going.

When the *Live-Stock Indicator* announced in a recent issue its convictions that the "coming steer" would be hornless—and correspondingly harmless, might have been added—it had no information that any of its western friends outside of the breeders of polled cattle had been putting the idea in practice among their calves.

But now comes Mr. P. R., of Kansas College, who is known as one of the most staunch short-horn men in the west, and announces that he began dehorning his calves last April, and has now about seventy head of mulley high-grade short-horns.

Mr. K. was in the city last week, and gave as his reasons for the practice that while in his experience with horned cattle, though no serious accidents had ever been caused by them, it invariably happened that the younger and weaker animals had to be separated from the herd in order to obtain their rightful food.

And this from the *United States Dairyman:*

DEHORNED, HAPPY CATTLE.

Mr. S. West, of Boone county, Indiana, writes us that he has made a grand improvement of his Duke and Bates breeds of cattle, by taking off their horns. He says their wicked dispositions went with their horns. Besides becoming more gentle, he finds he can put them in the barn like sheep, and a much greater number will eat from the same trough, and they have no fear of one another. They do not seem to suffer any ill effects from the dehorning, as they produce the same amount of milk and butter as before.

And this is from a Pharisee:

The lull of the winter season has brought the annual dehorning question on the scene again. This year it is full of scientific plans and methods that sound pretty difficult and somewhat dangerous. I cannot at all understand why wholesale dehorning is necessary or advisable. It is to my eyes very disfiguring, and I have not found horns any disadvantage in a rather long career of handling both beef and thoroughbred cattle. Sometimes a cow is bossy and her spirit has to be curbed. But is it requisite to visit her sins upon the whole race? Now, I have found that simply to saw off the tips of the horns of a bossy animal suffices to curb its spirit. Among short-horns we find such a large proportion of incurving horns that their owners can never be at all dangerous to anything, and this character of horn is always a great ornament. To sacrifice it in calf-hood, before it has shown how it will grow, appears to me to be a piece of unnecessary vandalism, on feeding cattle almost as much as breeding stock.

Capital growler, ain't he?

And now comes another. This time a regular old rip-snifter. This man is a knowist, the editor of the so-called "*Short-Horn Journal*," of Kentucky. Listen, while he speaks:

"THE DISHORNING CRAZE."

Some of our constituents have requested that we attack this dishorning foolishness that interested persons are trying to make a craze to help hornless breeds, and cause the unwary to buy them on that plea. The whole thing is too insignificant for us to dignify with an attack. One to read the articles would think that the beautiful horns that so much adorn the wearers were formidable weapons of destruction, more to be feared than the catapults of the ancients or cannon of modern times. That the poor, offenseless bull with horns is to be likened to the arch enemy of mankind who goes about seeking whom he may devour. From the amount of talk about the lack of horns we should judge that this deficiency was their greatest if not their only recommendation to the American people. Why don't they talk about their milking qualities, that would recommend them to the farmer and agriculturist? Simply because they are woefully deficient in that important part of the beef animal. Why don't they bring forward the beef qualities of their favorites? Because they know they cannot compete with the short-horn in that. Now as a last resort they make a great "bugaboo" of the *horn* question, and spin long yarns about the *danger* of horned cattle to draw the attention of beginners to them. They remind us of the fox that accidentally lost his tail and went around among his acquaintances trying to persuade them that the loss was an advantage; that it added to his personal beauty, and that he was better off without a tail than with one, and urging that the other foxes cut off their tails to be in the fashion. The other foxes saw through the thin artifice and told him they were satisfied with their tails and would keep them.

The short-horn men are perfectly satisfied with the horns with which their cattle are adorned. They know, too, that there is not so much danger in the thrust of the horn as in the butt of the head. All who have had experience know that a blow on the abdomen of a cow is more apt to make her lose her calf than a punch. The horns scratch off the hair, possibly.

Verily ye are *the* people, and wisdom shall die with you. The "dishorning craze." Dishorning is good. It is perfectly safe to assume that any man who spells it with an "S" also spells his I with a "hi" and "'ails from hould Hingland." Anyone disposed to treat the language fairly will at once admit that there is no need of two consonants in the word, and that to behead and to dehorn are alike euphonious and grammatical in construction.

At the outset of this question it never occurred to the short-horn men that dehorning might become an essential to their business; but it is, nevertheless. A race of short-horns that are without horns are the only cattle that can hold the fight as against the black cattle. In the opinion of the writer the farmers will have mulley cattle somehow, and the idiot who penned the above drizzle may live to see the truth of my proposition if Providence should spare his unprofitable life for ten years.

A Novel Objection to Dehorning.

Mr. John Boyd, the Jersey breeder, in a communication to *Hoard's Dairyman*, takes issue with Professor Henry on the subject of dehorning, especially of dehorning Jersey bulls, which he objects to, not on the ground of cruelty, but because he believes it will destroy the usefulness of the animal. His theory, which he admits is as yet unsupported by facts, is that by depriving the animal of his means of offence and defence, and breaking down his courage so that, as admitted by Professor Henry, his dehorned six-year-old bull was after a short tussle mastered by a two-year-old yet bearing his horns, there is danger that his prepotency will be destroyed, so that he will no longer impress his characteristics upon his offspring, and thus a dehorned bull of a noted butter line of descent will become no more valuable as a getter of butter cows than an ordinary bull of the same breed. Further on in the article he declares as another ground of objection to dehorning, that the "hornless cattle are probably the worst fighters in the world, and actually do more damage to one another than those furnished with the weapons nature gives them." This declaration is certainly open to question. But admitting its correctness, it naturally follows that the dehorned bull, instead of remaining a broken spirited animal, speedily becomes a better fighter than ever, so that instead of losing his prepotency, this quality would be increased.

As I have written in the *Review*—"There is nothing to the point made in the *Dairyman*." Old Dauphin 20th has given us three crops of calves since being dehorned. He is seven years old, and is as prepotent and as good a getter as before being dehorned.

The *Breeders' Gazette* at Chicago is *the* live stock journal of the Northwest. It is king of the crowd, and it was a great surprise to the writer to find its Mr. Dickinson reporting the Wisconsin address to the length of several columns. It is tony, perhaps, and feels its oats (ads), but it has been very fair in this discussion, riding no high horse at all.

Here comes a screamer from the North Pole, away up in Dakota:

To-day I cut off four sets horns, and the old cows did not care much, but, oh, do tell me quick. I cut off a four-year-old steer's horns (half Jersey blood), and oh! Oh!—I guess he will die. He bleeds awful, say six hours, a fine stream, squirting four feet. We are now trying to stop the blood, and don't know how. Now write me a private letter, and tell about such cases. The cows hardly bleed any, but think of this steer bleeding after nine hours, and still at it. Do they ever die? How to stop the blood? I certainly shall not dare to cut off any more till I hear from you. He bleeds on the right stub, not on the left. How much blood has a steer, and how long will it take a single fine stream to bleed an animal to death? Do answer quick.

COMMENTS.

All that ails this steer is that, first, he was hot when dehorned; second, the horns were not removed; stubs were left, which of course bleed and your fussing tends to prolong the bleeding.

The printed directions which go with the tools, will, if followed, save such experiences.

The *Farm, Stock and Home*, of Minneapolis, Minn., says of dehorning:

We advise our readers to "go and do likewise." We have heard from and conversed with a great many who have dehorned their stock and have never heard word of regret, nor an intimation of any ill effects; on the contrary the expressions were uniformly those of satisfaction and delight.

MINNEAPOLIS, MINN., March 18, 1887.

H. H. HAAFF, ESQ.

Dear Sir:—As a reader of my paper you cannot be in any doubt concerning my opinion of the beneficial effects of dehorning. I have published many letters from those of my readers who have dehorned their cattle, those who were inexperienced, who never saw the work done, yet their success was so complete, and their satisfaction over the results so entire, that they would hasten to assure.

NEW YORK, IOWA, 1882.

FRIEND HAAFF.

I have been dehorning for some time. When I commenced, prejudice was largely against dehorning, but it has given way, and dehorning is very popular here. I have dehorned old bulls and young ones, cows of all ages, in all kinds of weather and in every condition. Many calls to see our herd of polled short horned cattle. In all our experience we never had an accident, nor have we seen any bad results. Would advise all to dehorn their cattle and save shed-room, feed and accidents. D. M. CLARK.

No better authority in Iowa on farm topics than Mr. Clark.

The following letter from Professor Henry shows the manliness of the man. When he believes in a thing he knows it, and is not afraid to let his light shine. Contrast his course with that of Professor " day after to " Morrow at our hermaphrodite institution at Champaign. Over a year ago Professor Morrow heard all about dehorning at Princeton, Ill. He took notes at the institute then of my two hours' talk. Asked questions and seemed interested; made no objections whatever, and yet I have to hear of the first utterance since from his mouth. But with Professor Henry how different. He has no politics in his. Fears nothing. Cares only for truth. Is not a political professor. Teaches and talks what he believes. He got right up before seven hundred people and said out and out, " Dehorning is right, and you don't want to let your prejudice overbalance your judgment." The one is a man. The other I deem a milk sop. The one has a big agricultural department, while the President of the Board of Agriculture of Illinois says he was down to Champaign, and the other had five pupils taking an agricultural course. Comment is unnecessary.

PROFESSOR HENRY SAYS:

Dear Mr. Haaff:—I have just one point I wish to make in this matter of dehorning cattle. In spite of all the talk against it, based upon every conceivable assumption, I have yet to learn of the first person saying a word against the practice *who knew anything about it by direct observation.* On the other hand

I have yet to hear of the first person who practiced dehorning and was not pleased and satisfied with its workings. Let those who are so loud with protests bring evidence and not words, and maybe the farmers will listen more attentively to them; until they do, dehorning goes on.

W. A. HENRY,
Director Expl. Station, Madison.

Here comes an unsolicited testimonial, and he, too, is not afraid :

DUNLAP, IOWA, March 30, 1887.

H. H. HAAFF.

Dear Sir:—Your name will ever be remembered by the more humane class of stockmen as that of one who has, through great tribulation, done more to alleviate the suffering of domestic animals than any other man of this generation. Ten years will not elapse until laws will be passed prohibiting cattle with horns from running at large upon our highways, and more especially in our villages where the lives of our children are constantly endangered by those useless appendages. Trusting that you will wear with grace the honor to which you are entitled, as with courage you bore persecution in the past, is the sincere wish of yours, etc.

B. F. ROBERTS.

Another down-easter writes: "I had to keep my cattle in five different lots before dehorning them, but afterwards they were altogether, and not only did better, but it was not much more trouble to tend them all than one of the five lots before." Another is so pleased to think that, "as the animal has no apprehension of what is going to be done, the actual suffering is only momentary, and there is no dreading the operation."

My boys laugh at the matter of cruelty, and make a lot of mulleys right shortly after driving strange cattle home. A little practice and plenty of help to draw up the head, using two rings, as explained in the "printed directions," is all that is needed.

No thanks to the Humane Society that collects fines and puts them into its own pocket, and no thanks to the old fossil Board of Agriculture—the self-propagating board that is looking out for paps and political plunder, and what the *Tribune* calls "self-perpetuating." For all of this tribe of old barnacles dehorning will win, and the day is not very remote when horned cattle will be the exception.

Dehorning Cattle—Full Directions.

Turn the cut sidewise to look at it. Study the loops, and note before using how easily you can unlimber the animal's head. How easily you can haul up the rope taut. Use a good five-eighths manila rope, fifteen feet long, and braid two three-inch rings into one end. The stanchion is simply an old-fashioned stanchion, only it is five feet up and down in the clear, that is between sill and top rail. Notice in the cut that the head is only partly drawn up. The cut should show a man standing in front and raising the head, and not less than two men at work at the rope. Raise and draw up very tight. Bind the rope over the top rail and nose again and through the second ring to keep it in place. Now get the saw, take your knife and shave back the hair on top so that the saw blade will not clog. You will now begin to see why Mr. Haaff's saw is better than any other for this purpose. You can cut any turn and to any point with it, which is very necessary. Some knowing critter writes the *Breeders' Gazette*, and the *Farm, Field and Stockman* copies it as follows: " Cut three-quarters of an inch back into the hair." Well, all I have to say is don't do it. If you do, you will have a lot of heads that won't heal up the bone, that's all. For cattle three years old, cut into the edge of the matrix which you expsoe with your knife. On older cattle don't cut into the flesh any more than to make it sore, so that it will heal and granulate properly. On two-year-olds cut a quarter of an inch into the matrix, and in yearlings cut half an inch. A little practice and you will get easy and like the job, for it is a positive kindness to the animal. The party above named has, it is safe to say, read my book and struck out alone, and like the man who uses pruning shears, he will have to live and learn.

Dehorning Calves.

This is tedious, severe, and tiresome to man and beast. It cannot be done rightly without using the gouge, and once you have tried you will see the point. It needs four good men to dehorn calves. Throw the calf onto its back; two men hold the legs, one man the head, and one operates. Now, ready! Well, take a sharp knife and cut one-half inch deep, clear around the young horn, now use the gouge and you will lift out a horn every time. Examine the embryo when removed from the gouge and see for yourself. Now try it on with a knife alone. Pick away a piece at a time, and lo! when the thing is done and healed up there appears a Nanny horn, a nub on one side and a scale on the other. Reason? You need a tool adapted to the purpose so as to lift the whole embryo out bodily; any part left lives and will grow. I may here say that I have had an unlimited amount of trouble to get my tools made to suit me. I have made enemies of the manufacturers because I was so particular, and enemies of some who "order and don't receive," but all that is over now. My tools are malleable iron, will stand rough work, and dehorning is no child's-play. When it comes to be understood all will agree that I have produced tools that are adapted to the business, and a saw that is better for the kitchen than any now made. I am charged with seeking to make money, etc., of course. That is the "fly in the ointment," and we wouldn't be in this world if those flings were lacking. The tools are O. K., and cheap, and that is enough to say.

TOOLS TO BE USED.

In very many cases, parties who read an account of dehorning cattle in some local or secular paper, jump at the conclusion that what others have done they themselves can do, and so they proceed to remove the horns from their cattle with either a stiff-backed saw, or a butcher's saw, and after a time they are apt to write the author a letter much like the following:

CALDWELLS, WIS., Feb. 5.

H. H. HAAFF, ESQ.

Dear Sir:—Please send full set dehorning tools by express. I have not done much of this business; I trimmed nine head for one of my neighbors, but some of them don't thrive—they seem to be running too much at the head. What is the matter? Yours respectfully,

J. M. SMITH.

DUNLAP, IOWA, Feb. 15.

Dear Sir:—We are having considerable inquiry in regard to your dehorning tools; please quote your prices. We have been selling hog saws, butcher saws, etc., but they don't find them satisfactory. Yours etc.,

MOORE & CO.

YUBA CITY, CAL., Feb. 10.

Dear Sir:—I have just been reading your work on dehorning cattle. I am very much interested in the subject, and as soon as I can get your tools I shall try my hand on 160 head of cows, young cattle, then calves. I have tried a few with a knife and a common saw, following your directions, and I see the need of better tools. I remain, yours truly,

B. F. WALTON.

I give these three letters, taken at random from the pile received this day from all over the country, to show how others feel on the subject of proper tools after having tried tools not adapted to the business. I can give many such letters—in fact, I receive them every day. As I have explained, the saw to be used in dehorning cattle must be of such a character that while the hair is shoved back from the horn with the left hand, the operator can place the blade of the saw in the right hand back under the hair a little, upon the matrix, and so guide it as he strikes the proper point at the base of the horn above the ear, and then withdrawing it if need be and putting the saw under the horn, make a short cut to meet the first cut; or, if the head of the animal will admit it, continuing the first until the horn is cut off, and then using the knife part at the base of the blade to clip the hanging hide, and in that way prevent mutilation of the head. This may seem like repetition; but as this book is bought by the readers for the principal purpose of learning how to dehorn cattle, no apology is needed on account of repetition. Let me reiterate what I have already said: use a proper saw to dehorn older cattle, and don't attempt to dehorn a calf without using gouge and outcutter. I notice, in a recent issue of the *Farm and Home*, some one used his common saw, and then " daubed the hair with tar and stuck it together over the stub horns." This party expresses himself as very much pleased. In about six months from this time he will be very much displeased; for aside from having a lot of very sore heads, he will, during the year, discover himself in possession of a lot of cattle badly disfigured by stub horns, and in less than a year from now he will write me for proper tools and this book.

Mr. J. M. Davison, of Belle Plain, understands the matter; for, having discarded the ordinary meat and carpenter's saw, he writes : " I see the advantage of leaving no horns after the operation, and all I now ask is to get one job in the neighborhood, and the rest are soon converted to the system. I sometimes break a saw-blade ; send me a dollar's worth, so that I may be prepared." People who use meat and carpenter's saws, will inevitably sometimes break them, and in so doing they destroy a valuable tool, beside improperly performing the operation. When the young doctors from the veterinary college and the dealers in veterinary instruments find it advisable to purchase these tools, and when the mere nominal price at which I provide them is considered, there is no excuse for any man having a lot of sick or stub-horned cattle, made by the use of improper tools.

BLEEDING.

In dehorning cattle, bleeding or hemorrhage may occur from several causes, but is first and most generally caused from bruising, and too much care cannot be exercised in handling cattle to avoid bruising the head or, for that matter, any part of the animal; and this is an additional reason why

I prefer my new mode of securing the head to my former way of using the stanchion, for it must be apparent to the most casual observer that, after the Jewel is once firmly placed upon the neck of the animal and the lever or hand-spike is inserted, there is not much danger that the animal will bruise itself.

Again, a second cause that promotes hemorrhage is from the animal being over-heated or in an excited state, and this can only be prevented by care. For instance, one gentleman wrote me that he dehorned a herd of cattle for a neighbor. He drove to his place, because he had the tools and equipment for performing the operation. He said the neighbor's cattle were driven about three miles, then dehorned and carefully driven back; they seemed to be all right until the next day, when one was discovered bleeding, and in spite of all they could do the animal bled to death. The operator, very much chagrined that he should lose an animal, wrote me asking the reason. The reason is apparent: the brutes were necessarily warmed up by driving three miles; they were more or less excited by being in a strange yard, amid new surroundings, strange faces, possibly some dogs to help (you know dogs have such a soothing effect on cattle); probably the man and his help were laboring under a little excitement—excitement is incident to such an operation when not understood thoroughly; the animals were dehorned, turned loose to go home, the excitement was kept up necessarily by reason of their anxiety to reach home. Cattle are like children, they hate to be away from home over night. Very likely somewhere in the rush this animal got its head bruised; weakened by the reaction, there was not contractile force enough left to the inner coating of the artery to collapse sufficiently so as that the blood would coagulate and stop the bleeding.

Since writing the above I have received the following letter:

HASTINGS, NEB., March 8, 1888.

MR. H. H. HAAFF.

Dear Sir:—I received your tools on February 22, and on the 23d went to work to dehorn my herd of thirty-two head. I had the best of luck with all except my short-horn bull, General Newham, No. 80,953, which was two years old the 17th of last October. After taking off his horns he bled quite free out of his right horn, but thought it nothing strange, as some of the cows bled just as much at the time, or more, so I paid no attention to it until the next morning. I found he had bled until he was getting weak; then I took him up and tried to stop it by various ways, first by applying cotton and tar, and that would not stop it; then I singed the artery and that did not stop it; then I stitched it up and bound bran and shorts on it, and that stopped the blood, but the bull was too far gone, he died after living about fifty-two hours. He did not bleed continuously, but took spells of bleeding about every six hours, and when I thought I had it stopped it would start again. Now, I would like to know if you can give any reason why he took spells of bleeding, and your remedy to stop blood in case we have any more trouble. I have let my brother, George Way, have the tools and he has been dehorning; has dehorned about seventy-five head so far with the best of results, except the above case. I would not have the horns back on my cattle again for three dollars per head and compel me to keep them on. Yours truly,

J. S. WAY.

The burning was bad, very bad. Dry flour at an early stage of the trouble would have saved the animal. Here is a record of one of God's noblemen. Instead of whining, as Guppy did, because Webster, of Marysville, Kas., happened to leave a few stub horns on his cattle's heads, and rushing into the papers to denounce Haaff, this man says "I wouldn't have the horns back again for $3 per head," and all this with a dead imported bull. Keep your animals cool and quiet before and after operating. Don't excite them.

I am not surprised that death followed in this case, because the patient was not found early enough to stop the bleeding by artificial means. The bleeding by spells was partly owing to exercise and partly to the natural supply of newly-made blood which the artery could not retain.

There is one other reason that may be given for an unexpected hemorrhage in certain cases, and that is from disease through a low condition of the system; some blood disorder, or more likely than any other, a frozen horn, followed by the death of the membrane and subsequent disease of the matrix. In this case the cutting of the part is attended with more or less hemorrhage, but I have never known any serious results to follow if the animals were quiet and cool when operated upon. I have myself dehorned cattle with heads so diseased, at the base of the matrix, that it was an operation not only offensive in the extreme, by reason of the suppuration of the parts, but necessarily attended with much hemorrhage; in fact, I remember one case in which a large steer had had his horn broken near the base, and I was obliged to make two cuts to remove the part, it was so terribly swollen and inflamed. In this case I advise the use of some dry astringent, and, if it can be obtained, I think the common dry puff ball of the fields an excellent application; if not readily obtained use flour; make a continued application of dry flour, being careful to somehow confine the animal or handle so as to prevent exciting it or otherwise.

SORE HEADS.

The previous chapter on bleeding necessarily includes much that might be said here, and which it is not necessary to repeat.

It will be understood, from reading this book, that I insist that poor tools must not be used in dehorning cattle; so also the stanchion and the chute must be properly built as herein directed.

In the case of frozen horns, or of horns that have been broken and are therefore remaining as sore stubs or left hanging from the matrix, I recommend in cutting that the operator cut a little deeper than in the case of heads not diseased.

To dehorn one of the case just named, and leave a stub horn after the operation, is to simply run the risk of losing the animal, and in any event of prolonging its misery; for if the operator fails to remove, nature will be likely to remove by a process of inflammation and suppuration with atrophy of the parts.

It sometimes happens that cattle which are dehorned late in the spring or early summer become fly-blowed, and the frontal sinuses or cavities in the head fill with maggots. A careful study of the chapter on the bones of the head will have taught the reader that there is no particular danger, in that event, to the life or health of the animal, but of course the maggots should be removed. I have observed that these maggoty cases are more likely to occur where, from some reason or other, there has been a considerable flow of blood through the opening into the frontal sinuses. Simply take a pine stick and poke the maggots out of the head; don't use tar or turpentine as an application, but take of ordinary axle grease, say half tea-cupful and mix with it half a dozen drops of carbolic acid diluted with ten times its bulk, say sixty drops, of sweet oil or unsalted butter; mix these with the axle grease, saturate cotton thoroughly with the mixture and fill the hole with the cotton, lying very loose and not packed into the orifice; daub the edge of the hole next the sore all around and the exposed bone with a little of this mixture. If any maggots happen to remain they will crawl out and drop to the ground—as indeed they all would if given a chance, for the grub must find the ground to fulfil its destiny.

I am frequently asked if dehorned cattle will receive injury by being allowed to run to loose stacks of straw or hay after the operation has been performed, and if there is no danger of chaff and other foreign substances getting into the frontal sinuses and making trouble, and some of those foreign importations that we have in this country, who are more snobs than veterinary surgeons, inveigh against the practice of dehorning (which, by the way, I notice they all call dishorning instead of dehorning), and declare that injury will surely follow the presence of foreign substances. I simply wish to remark here that I would not hesitate to fill the head of an animal properly dehorned full of oat chaff, and I don't believe, in fact I know it would not, militate against the health of the animal or the process of healing, with this exception, that if the chaff became wet, in very cold weather, it would tend to reduce the temperature of the parts too much, and in that way delay the process of healing; for it must be understood here that an animal properly dehorned in cold weather has had the very best possible operation performed when the head is left so that the hair, as near as may be, covers the opening loosely and the hollows in the head are left free to slowly emit the warm air. This is Nature's salve, and is the only help which the flow of serum needs to effect a perfect cure.

CHOKING.

A great danger in performing the operation of dehorning is, by reason of the brute's head being closely confined either in the stanchion or in the chute, the cross-bar or side of the stanchion may come in contact with the windpipe of the animal and prevent his breathing. The skillful operator in dehorning will make so quick a job of it that sufficient time can never elapse to seriously choke the animal before it is liberated. But it sometimes happens that the animal in dehorning gets caught or confined, and in any event it will be the business of the successful operator, by occasionally passing the fingers of his left hand over the nose of the animal, and know if it is breathing or not. In using the chute, possibility of choking is prevented by drawing back the first bar under the animal's neck, after the animal has been secured by the Jewel and the bull-dog; liability to choking is largely removed by reason of the slanting form of my new chute, as that form of chute prevents the possibility of the animal casting itself. Cattle choke much easier than humans, and bulls that are plethoric in body much sooner than ordinary stock-cattle or milk cows. Watch your animal, and in that way prevent the possibility of choking. The only animal that ever died on my hands, or under my administration, was a bull that I foolishly allowed to be dehorned in an ordinary cattle chute, built on the ground, with a stanchion to secure his head at one end. As it happened this animal, as was apparent from the appearance of his blood, was out of condition; but had there been a proper place in which to operate on him, I am satisfied that the result would not have been fatal; as it was, he was simply choked over the slanting side of the open stanchion because he had so cast himself in the chute that it was impossible to quickly release him. They were a company of Hollanders, and in speaking of it afterwards, the owner got off a good one on the author; said he: "I think my bull was talked to death." But the author thinks they were not good judges, since the multiplicity of glasses of beer which they absorbed during my visit and (which I never take) led them to mistake their own verbosity for mine.

DEHORNING.

Why recommend it? Why do you recommend dehorning without qualification as to age, or place, or condition? This is a question recently asked me by some who are skeptical as to the merits of the practice.

The ordinary reasons for dehorning are herein coupled together; but since the question is asked, I ask in reply: Why not dehorn? Really, now, my reader, will you pause for a moment and ask yourself this question: Of what possible earthly use are horns at any time, or under any possible circumstances or condition that you can think of? Do you tell me that in the case of cattle attacked by wolves or mountain lions, or even by panthers, that horns would be of any possible use? Do you say that the cow might protect her calf from such attack? I say no! I was of that opinion myself; but some months spent in Texas, in the Indian Territory, and at ranches on the plains elsewhere, satisfies me that that is a mistaken view of the subject. I grant you that wolves and lions sometimes kill a good many calves, and occasionally some older cattle; but I challenge you to find a ranchman who will say that horns on the cow will save the calf. At the Gohram ranch, of 145,000 acres, on the Cimaron river, I find, as elsewhere, that they lost as many calves by wolves, both coyotes and also the large gray kind, and there the cow-boys told me that it made no difference whether the cow had horns or not—that wolves hunt in couples, and that one will pay his respects to the mother while the other steals the calf; and so true is it that cattle cannot protect themselves, that these ranchmen scatter poisoned food, put up in various shapes, at times when their calves are less liable to be killed, in the hope of thus decimating the ranks of these marauders; and at the Gohram ranch, on a short ride which I took with some of the cow-boys, we saw the remains of several calves recently slaughtered by the wolves, although at that ranch they had a pack of staghounds, the only kind of dog known that can successfully attack wolves, and even the presence of these dogs was so unsatisfactory that the men told me that the intention was to rid the place of all she-cattle, stop raising cattle, and buy steers of other ranches.

If horns are not an unqualified and unmitigated nuisance, then the men who are engaged in the business most extensively are not qualified witnesses; for at Ames, Neb., where they are feeding at this writing 6,000 steers, the superintendent, Mr. Allen, told me that he could well understand that if they could get rid of those useless protuberances it would be many dollars in their pocket in driving, in handling, in feeding, and in shipping.

I have within a day or two received a letter from Hon. R. H. Whiting, Congressman of the Peoria District, at whose ranches in Kansas I taught the art, and who has herds of thorough-bred short-horned cattle, and who believes and knows that horns are a source of great loss every year. He gives the result of an experiment tried by Mr. A. Prout, of Severy, Kas., who recently sent a load of dehorned cows to market, and obtained at the packers 20 cents per hundred weight above horned cows of same weight, flesh and fat; the packers paid more for the dehorned than for the horned cows, " because they were free from bruises." This netted Mr. Prout something like $2 per head for his experiment; and besides this, they had required less feed and far less shed room, and "Mr. Prout is convinced of the merits of dehorning."

When to Dehorn.

As a rule, cattle may be dehorned at any time of the year. I have been accustomed to say any time except fly-time, and when the thermometer is at zero; but I am satisfied that cold weather is a benefit rather than an objection if proper care be taken of the animals as to food and water with a comfortable shed and free access at all times. Dairy cows or cows that are in calf should not have the hind bar placed under them during the operation. Fattening cattle may be turned loose after the operation at will. In the case of yearlings the head should be shaped, so as to give it a more conical appearance. This will be better understood by reference to the figures [see *ante*]. Calves should be handled with care; the operation is a severer one on them than on older cattle, but by the exercise of ordinary care both the operation of dehorning and castration may be performed at the same time. Calves and yearlings will necessarily bleed more than older cattle, for the reason as explained that the membrane surrounding the bone horn is larger during the early life of the animal, owing to the rapidity of the growth of the horn.

Dehorning with shears should never be resorted to; the operation of crushing the parts with ordinary hedge shears or other instrument of that kind cannot be too strongly deprecated. So also the operation of burning either calves or older cattle with a hot iron is barbarous and cruel in the extreme, as they destroy part of the matrix which would materially assist in the healing, and compel nature to rot out the burned substances, restoring it in part again before the process of healing can be completed, and also by weakening the arteries this practice renders bleeding more liable to occur. So also "putting something on" simply irritates the parts and retards the process of healing. I have explained before, but I will repeat it: immediately after the operation is performed and as soon as bleeding has ceased, nature pores out on the parts a serum which is her salve—nature's own restorative—and the process of healing begins simultaneously with the flow of serum.

In the case of refractory young bulls which are to be kept for breeding, or heifers needed for the dairy or home supply of milk, I have found it beneficial to dehorn such animals in a stanchion and leave them confined for days—in fact until, by careful handling from day to day, they learn to omit their refractory habits and become docile and tractable. I throw this out as a suggestion for the wise farmer, who knows that personal attention to young animals is what makes them gentle.

SOME COMMENTS ON THE HEAD.

The bovine head is most singularly made up. It is properly divided into two parts. We are speaking of course of that part of the head which is connected with the horn, what we would call that part connected with the upper jaw. The dividing line is down the center of the forehead, and is called the suture; the two parts so divided by the suture are themselves composed of many bones but all united as one whole. An examination of the skull of the ox will show the observer how wonderful the construction is. The brain lies well protected down under and next to the parietal bone, while the parietal bone itself is completely covered by the frontal bone and its various parts and ramifications. Take the skull of an ox and divide it at the suture, each part seems to be a perfect bone, through both the frontal bone and the parietal wall beneath into the cavity that holds the brain; it will be readily perceived that in striking a blow upon the ox head the most vulnerable point is precisely in the center of the forehead, where two lines drawn from the eye to the base of the horn would cross each other. At that point in the grown animal the two bones are usually not much over a half inch in thickness, both of them. That is the point at which the hunter or butcher aims when he shoots the animal in the field. That is the point where the butcher at the slaughter-house was wont to strike the animal with his hammer to kill it. Now a moment's reflection will show that a blow on the end of a horn will tend to separate the two parts of the head at the suture; and of course, as these parts simply touch each other and are held in place by the hide and by various ligaments provided to join them together, a blow on the horn would so spring the parts as to produce great agony to the animal.

In this connection it may not be amiss to mention the fact, that in rounding up cattle on the plains or in staying them when stampeded, they suffer loss and great agony by the way in which the rounding-up is done. The cattle are run on a circle which closes smaller and smaller, until finally in the center the cattle themselves, overborne by outside pressure, will stand erect on their hind feet; and as they are by the outside rings more and more crowded against each other, the horns are the first points that suffer, and they sometimes break by scores.

In whatever way we may view the matter, or at whatever point we may consider it, there seems to be no exception to the general rule that horns are a nuisance; a source of danger and of damage to the brute itself, and an injury and loss to the owner and to all who have to do with cattle. It would seem to be unnecessary to call attention to the great loss which horns occasion in shipping cattle; in fact, from the time of the grand round-up on the plains of beeves for market, from the time of the drive on the ranch to the corral or stock yard, from the time that the cattle leave the home farm, in whatever way handled, the horns seem to be like Uncle Joshua's corns, always in the way and forever getting hurt. What thrilling stories some men can tell of their experience in getting into close cars to relieve the downed cattle; how many of the cattle themselves die on the way to market or are rendered comparatively worthless by reason of bruises and horn thrusts received in transit; what suffering and untold agony the poor dumb brutes undergo; brought up on the prairie, farm or plain, accustomed to perfect freedom in their movements and ways, they are now thrust into a crowded yard, from thence into a worse crowded car, and finally between themselves and their inhuman masters, they know nothing but misery from the moment they leave the field or the feed yard, until the discharged bullet tells that all is over, and life with them is ended. I am sure that the inhuman brutes who prod them with their sticks and iron pins, who dog them on the road, who beat them while they are being carred, who "holler" at them when they use their horns too freely on their fellows, who hound them on their way from the cars to the slaughter-house, and who finally end their misery and give peace to their fevered bodies by the well-aimed bullet, through the top of the scull at the back of the horn—I am sure, if they paused to consider, they would at least with one voice declare, that to rid the poor brutes of their horns before leaving the farm or the ranch for the market, would be nothing but unqualified kindness to the animal; and it would seem as though the men who declaim against dehorning as cruel, who criticise those who practise this art as barbarous, and who inveigh against the practice without reason or justification, I am sure if they would each of them pause and think how much misery and suffering the animals inflict upon themselves, and how much unnecessary suffering their horns cause them in transit, they would stop their mouths and be dumb, or, if speaking, would admit that the practice of dehorning is a most humane one and ought to be adopted by all.

CRUELTY TO ANIMALS.

There are many persons who are converts to the practice of dehorning, and among them some editorial writers, but who in discussing the question of dehorning are wont to admit that it is cruel to dehorn cattle, that it is cruel to castrate, to spay or to brand; in fact, with them the rule seems to be, that to inflict pain is to be guilty of cruelty. This is a mistake, and I wish to enter my protest against anyone allowing that cruelty and the infliction of pain are synonymous terms. Cruelty is the infliction of *unnecessary* pain, and to dehorn an animal is not to be guilty of cruelty, because the pain inflicted is necessary, the pain inflicted is warranted by the results obtained; it is not cruel for the surgeon to amputate the limb when necessary to the welfare of the animal brute or human. It must be borne in mind that these brutes are a gift from God to man, that he is permitted to exercise his will upon them and for his own benefit, that by the Divine law as well as by human permission man may do his will upon them, provided always that the end shall justify the means used. This, then, constitutes the true rule: does the end justify the means? and when man finds that his cattle which were pugnacious and unruly, an injury to themselves and a loss to him, become by the simple act of dehorning quiet, submissive and infinitely more tractable, then man has the moral as well as the legal right to remove the horns from his cattle, provided always a way shall have been found for performing the operation so that there shall be little and only momentary pain to the animal, and no loss of life or resulting damage.

Enough has been said already on this branch of the subject; no one who is well informed will now contend that the fact that horns were put there constitutes any present valid reason why they should be kept there; no matter why " they were first put there," God placed them there—that is the answer, and God has by his holy law in Genesis, the 1st and 9th chapters, given man the perfect right to remove them. If Divine sanction or warrant is necessary, it is found in that chapter, wherein the Almighty Himself declares that he has handed these cattle over to man that man may do his will upon them. (See Gen. and Ps. 75.)

SHEDS FOR CATTLE.

In the little book " Haaff on Dehorning," I gave a chapter on this subject, in which I proceeded to recount some of the advantages to be obtained from having sheds built on a side hill. It is objected that on a steep side hill cattle cannot lie or stand with the same measure of comfort and repose to themselves that they do on level ground. This is a mistaken notion; cattle will always seek a dry, sunny place in cool weather, and they will never lie down where the ground is wet unless compelled so to do by force of circumstances. If the reader is at one with the author in his desire to make the few years of the animals' existence here on earth as comfortable as may be, then I beseech you to find a good steep side hill, somewhere facing to the south, if possible, and there erect your cattle sheds.

My plan of building a side hill shed is so cheap, so simple, so readily and easily done, that there is not the slightest excuse for any one who leaves his cattle unprotected without good, warm sheds to protect them in winter from the cold and in spring and in summer from the storms and heat. (See the cut.) Winter sheds, if in a locality where lumber is not too expensive, can be best built by the use of common sixteen-foot lumber and of posts long enough to be set into the ground, say a foot or a foot and a half, giving a clear space between the ground and the roof of seven or eight feet, just as suits the owner. Set such posts into the ground on a side hill, so that when set up they will be distant from each other, each one eight feet from center to center. Nail at the back on each side common fencing boards, giving them the slant of the side hill, and so nailing them that the back edge will give a smooth surface as may be with the other rows of boards on the other rows of posts; of course these rows of boards will run lengthwise of the hill and shed, so that when the roof boards are put on they can be nailed at top into the edges of those cross-boards. A hill-side should be chosen that is tolerably steep, something like an angle of, say 60°. [The cut not steep enough.] If there are depressions or holes in the side hill, use longer posts, as the case may be, so that the line of boards nailed at the sides of the posts at the top will be substantially straight throughout. This will give a good equal appearance to the roof of the shed when completed, and in looking at the side-hill shed when completed, the roof is about all one sees. It is immaterial whether the roof be built first, and then the sides be added, or whether the sides are first erected. I believe I would put on the roof first. Now to build the side and end of this shed we have only to nail boards on each side of the posts along the north side, and the same way along the east and the west ends, and as the boards are nailed on fill in with any light substance that is handy. In building such a shed on my farm, 160 feet long and 30 feet wide, I filled in between the boards with dry rotten manure, because I happened to have it handy, and it made a wonderfully warm shed. Board up along the south side of the shed with a single tier of boards, leaving an opening every 30 to 40 feet; it makes little difference which, for the cattle, being dehorned, will not interfere with each other in the large shed I have mentioned. Two large ten-foot tanks for water I had in this shed; they were lined and were 2½ feet deep; I kept them supplied by the windmill on the ground on the north side of the shed, and they occasioned me very little trouble in winter by freezing, but it is immaterial now whether the farmer has his tanks in the shed or outside; for if outside in the most exposed place, and the farmer has one of my cheap patent tank-heaters arranged as by me directed, he will always have warm water for his cattle to drink. Until I had such an arrangement for heating the water, a close shed like the one above described was the place for the water-tanks.

I recommend my readers to build their side-hill sheds fifteen feet wider than I built mine; if they will do that the necessity of cleaning the shed out in winter will be largely removed, and you will be astonished to see that the lower fifteen feet of that shed will during the course of the winter fill up with manure to the depth of three to four feet, so that you will be obliged to frequently clean the manure away from the entrance passage, which is best done by throwing it back away from the door on the inside or carting it away. Don't throw it outside to daub the cattle and be wasted. You will be equally surprised to find that your cattle will be always dry, and you will be glad that you took my advice and built your shed on a "steep side hill."

I do think that every man who reads this chapter gets more than ten times the cost of this book from this chapter alone. Feed your stock cattle once a day during winter, all they will eat of hay or whatever the food may be; give them free access to such a shed as I have described and to plenty of fresh warm water; and if your cattle enter the winter campaign in good order, my word for it, they will start the succeeding spring campaign in equally as good order and condition, and you will be surprised to find how closely those cattle, all of them, calves, underlings and all, will remain in that shed during the cold weather.

It was a wise provision of nature that gave the cow brute a stomach so built as to be able to carry food for many hours, and it is not a mark of good judgment on our part if we don't take advantage of this fact in wintering cattle.

I have said that there is no excuse for any one being without a cow shed in winter, and I mean it. Were I located on the plains of Kansas, or of Dakota, or of Nebraska, or of Montana, or of Texas, either I would find me a side hill, and I would find something for posts, and lacking boards I would lay poles along these posts and other poles crosswise, and on top of all I would have a cover of some kind of straw, or hay, or rubbish, or what not, and it is no particular matter either if it does leak some; it will fill the purpose intended in any event, because it will be comparatively easy to fix the upper sixteen feet so that they shall shed most or all of the rain. For outside, in such a case, lacking boards I would cut sod, have a sod side and ends—if possible a better outside than double boards are, even though filled in between. I repeat it for your benefit, my reader: if you expect to make any money in handling cattle, you want such sheds as I have described, and you should never allow yourself or your boys to be so miserably shiftless or lazy as to enter the winter campaign without being well provided in this respect. I don't want to hear men talk who have so little sense as to suppose that "cattle can become accustomed to weathering the storms of winter." It is not true—they never become accustomed; they may have to endure it, but if they have to it is because you neglect your duty, and are so far shiftless and improvident. I have no patience with a man who will not protect his cattle from the storm of winter and from the inclemency of the weather. Not long since a farmer wrote me saying he had no sheds—could he dehorn his cattle? and I said, " No, you ought not to do it, and you ought to be prosecuted for inhumanity if you do," and finally I said, "You ought not to be allowed to have any cattle at all;" and I add, in closing this chapter, that such men are a disgrace to the profession of farming and are guilty of gross sin against God and their beasts, to say nothing of the fact that they sin most egregiously against their own pockets and their own financial welfare. The contemplation of your many cattle running in the country during the winter foraging for food among the brown tops of the rustling weeds or the sun-dried ends of broken-down corn stalks, gives me such a feeling of misery and dreary want as makes me fear that we may some day be a nation of capitalists and serfs instead of a nation of intelligent and free-born American citizens. I urge you, gentle reader, not to permit your cattle to pass another winter without good ample warm sheds; and I know if you will take my advice, that when I next meet you, you will take me by the hand and say, "You are not only a public benefactor in teaching us how to dehorn our cattle, but you have done me great good in showing me how to keep my cattle fat during the frozen winter."

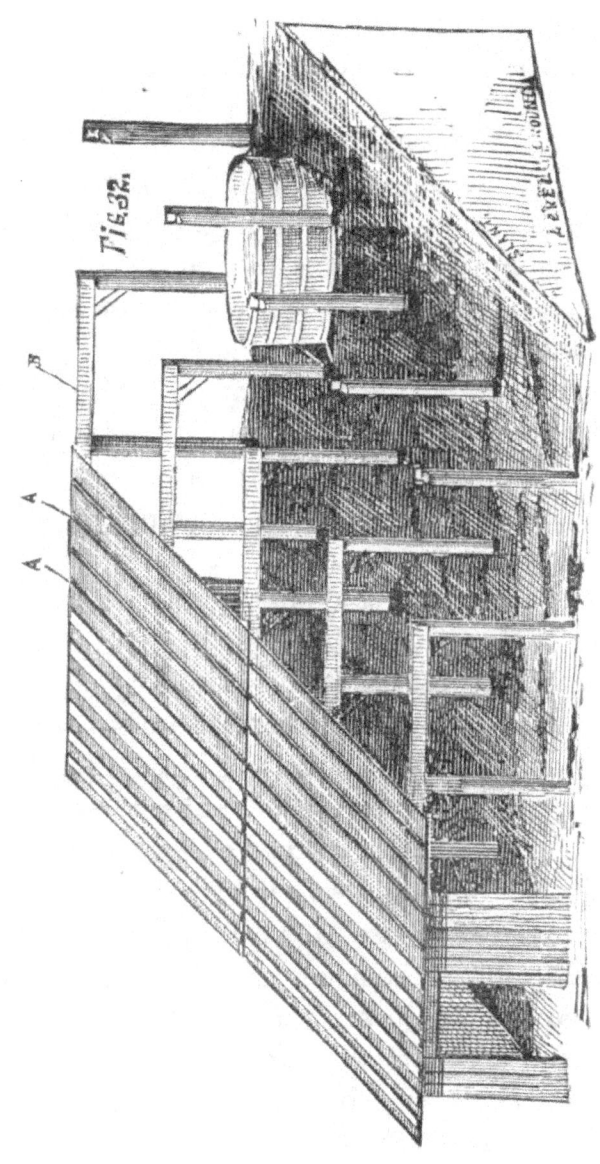
Fig. 32.

WARMING WATER FOR CATTLE.

I wish to call the especial attention of the farmers and stock men to my system and patent apparatus for warming water for cattle or for other purpose, and for warming swill or slops for hogs. A tank half-full of ice or swill barrel half-full of frozen swill, the one surrounded by cattle trying to drink and the other by hogs filling the air with piercing screams, while the negligent owner is striving to cut a hole in the ice, so that he may get at the frozen fluid, are both of them pitiful sights. See the poor horned brutes standing by, drinking the ice-cold liquid, opening their mouths in agony at the pain inflicted by the ice water, and stretching out their limbs behind and shaking them as in a veritable chill, or doubled up after being filled with the chilly stuff, giving the appearance of suffering from genuine ague fit! Look at the pigs—they have gulped down the half-frozen swill, and now see them go like a lot of inflated bladders, seeking their nests in the straw stacks or a pen somewhere, anywhere to get warm again! I never witness a sight of this kind that my heart don't go out to the poor brutes in about the same ratio as my indignation rises against the owner. Have you ever stopped to think, my neighbor, that no animal can long survive if the blood is lower than 98 Fahrenheit? Don't you understand that when you give an animal frozen food or drink, it detracts so much from the heat, that is, from the flesh of the animal, as is necessary to restore the equilibrium of the body and bring it all to 98°? It is possible for you to take pounds of flesh from your animals every day in the winter; and I tell you flat-footed that every time you let cattle drink frozen water and stand shivering and curled up in your sheds, you lose from one to five pounds of beef a day per head, and this is just as sure as that I have written it. If beef is worth $3 a hundred, and you have a hundred cattle, you have simply knocked $3 to $15 out of your pocket every day that your cattle fill up on frozen water; and if you will sit down and figure the number of days there are in the winter campaign, and if you have weighed your cattle when they entered the winter campaign and you now weigh them again—spring-poor, some of them perhaps "on the lift," some of them looking as though the crows had a mortgage on them and as though the mortgage ought to be foreclosed—you will understand, and you can figure and see that I am moderate in my calculations when I tell you that every such day costs you one hundred or more pounds of beef.

Now, I want to explain to you a little of the philosophy of my plan for heating water, so that you may understand it yourself. The full directions accompanying my "Tank-Heater" are: A simple roll of pipe, with the ends turned up like the letter U, run through the bottom of tank, the roll left underneath, as shown in the accompanying diagram; a cheap wooden box with the four sides remaining to give you the hot air chamber under the tank, a lamp containing a gallon of kerosene, and for 12 cents a day, and a half-hour of time, your cattle need not drink an ounce of cold water during the winter. My galvanized iron tank-heater is so arranged that by lighting the lamp and inserting the heater into the barrel of water or swill, the same may be heated to any degree in thirty minutes' time. This heater costs you one-half the price of the advertised heaters, and will give you the very best of satisfaction. It is so simple in construction and so easily handled that I know every farmer will fall in love with it on sight, and thank me for having given him something besides the horrid soft coal and wood heaters of the men who do their farming by proxy and play "city chap" while posing for our benefit.

A serious objection to one of them is that you can't set a stove into the side of a tank and keep it tight for any length of time, while an equally serious objection to the others is that the cattle will surely rub, rub on those side rails until they rub the whole thing loose. My heater has a corkscrew arrangement, and can be screwed to the bottom of the tank at any point. It is just the difference between men doing things who know nothing about them by practice and a man who is a practical farmer and cattle man and knows what you all need.

DOGS.

I approach the discussion of the advisability of allowing a dog on the farm with some hesitation, for the reason that the home prejudice is stronger on this line than on that of any domestic animal on the farm, and yet I have deemed it my duty to attack this stronghold of prejudice and do what I can to create a sentiment in favor of *the utter obliteration of the canine species*, and we may as well ask the question at the outset, as the question is propounded regarding horns, What earthly good is a dog on a farm, and what are dogs good for anyway? A dog on the farm costs more to feed than a hog; the price of a well-fed hog on the farm is at the present writing from $15 to $20. It is a low estimate when we allow that the food consumed by the worthless cur would fatten two hogs each year. Think of it, farmers; the worthless, good-for-nothing, yelling, yelping, snarling and racing cur costs you every year $30 to $40 in the item of food alone, enough to buy a pair of boots each for yourself and your five boys—enough to buy your good wife a silk dress, better, I venture to say, than she has worn since she married you; enough to buy your boys a good set of blacksmith tools—enough in two years to make them perfectly happy and you perfectly independent of either carpenter or blacksmith for all ordinary farm jobs;

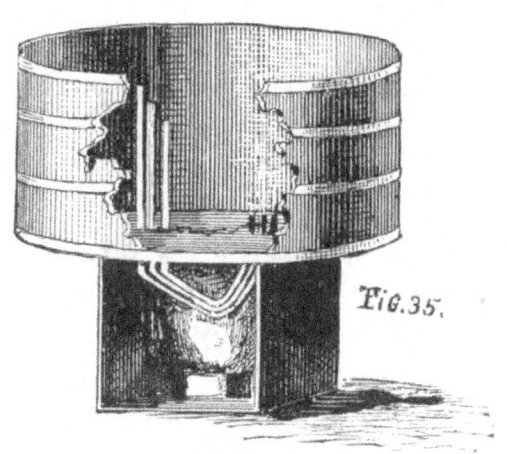
Fig. 35.

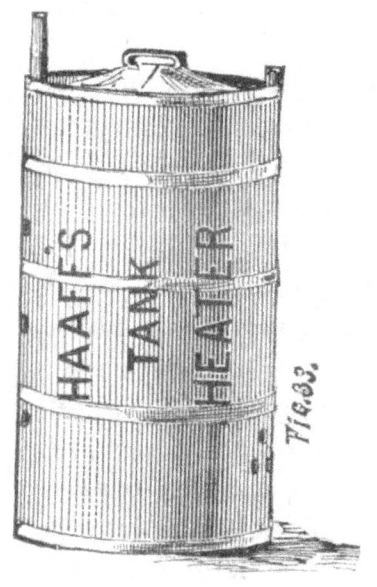
Fig. 33.

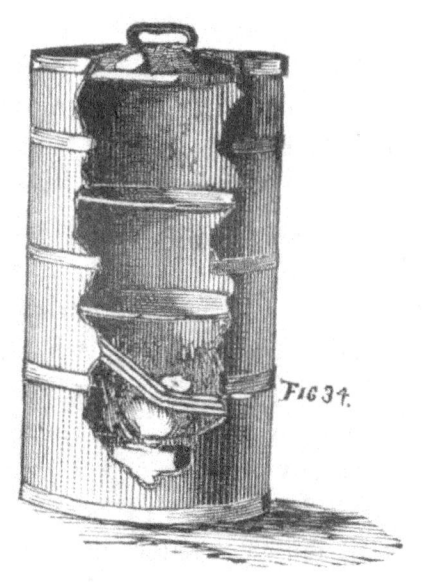
Fig. 34.

enough to place on your table twenty of the best papers printed in these United States; nor is this all. If the cost of the "purp" was alone in the item of food we might dismiss it, but this is the simplest item of the dog's cost; the chasing of cattle, the worrying of hogs, the destruction of sheep and fowls, and I dont care how good a dog your "Boos" is, you can't deny, but you must confess, that he is a nuisance so far as all stock on the place is concerned. It is a thing that I could never understand why a farmer wants a dog. I have never seen of the so-called shepherd dogs a dog that could well fill the place of a boy or girl in driving up the cows, and I have never seen where a boy was kept for that purpose an instance in which a dog was really needed. It is true, you may think he is an essential on the farm; so, too, you thought that horns were essential until you learned a better lesson; but if you will now seriously consider the matter and impartially weigh the evidence so far as your dog is concerned, I am sure that the discussion of the matter pro and con will lead to the inevitable conclusion that he ought to be "dehorned just back of his ears."

I have added for the benefit of this chapter a few cuts giving truthful illustrations of actual occurrences of every-day scenes of dog economy. I know what you will say in looking at the illustration of the dog worrying the cattle or worrying and biting the sheep—you will say: "My dog never does it." But now, good reader, you know that is not so; your dog is not guilty in that line perhaps to the extent that some other dogs are, but your dog is a dog nevertheless, and he is sure to use his dog proclivities when he has an opportunity and feels like it; nor do I believe there is a single farmer who can tell during any one day of the year what his dog is doing or where he is. I am not in this chapter proposing to enter into the discussion of the dog question so far as the loss of sheep is concerned, for it is simply a one-sided question, and it is to be answered in one way, and that is: "All dogs ought to go!" How can a farmer who keeps a sheep dog expect to complain justly of losses in that line from other people's dogs? the mere fact of his possession of the canine brute is a continuing sign or admission that his neighbors are justified in keeping any kind of dog brute they see fit; for you remember the story of the worthless dog on the farm; his owner was asked what he was good for. Was he good to drive cattle? No. Or hogs or sheep? No. Was he a good watch-dog? No. Well, was he good for coons? "Well," said the farmer, "I guess he must be good for coons, for he is good for nothing else." So far as the matter of dogs for purposes of watching are concerned but little needs to be said. Any thief or number of thieves who propose to raid the place will be very little deterred by the presence of a dog; they will see to it that the animal receives a dog button or a chunk of lead before they proceed on their thieving mission, and your dog, instead of being a hindrance in that respect, ofttimes proves a help. It takes but a moment to lull him into quiet with a bit of poisoned meat, and then as you fail to hear his bark you are lulled into a sense of security and are oblivious to sounds that would otherwise awake you from sleep and attract your attention. Is it in the matter of handling cattle? I have handled thousands of cattle every year for the past thirteen years, not only my own cattle but thousands of others of my neighbors' cattle brought to my pastures, and I know by actual experience and by the admission of the very men themselves who were accustomed to think that they could not drive cattle without the use of a dog, that I could handle even their cattle with a boy and a horse, or by the call of the voice far more readily and with infinitely better results to the cattle themselves, than where it was attempted to let the dog do the driving. I have taken such cattle on many occasions, myself alone, and called them from one field to another, possibly for a mile, by the voice alone, transferring them, oftentimes by the hundred, before my breakfast in the morning. I need not discuss or enlarge upon the injury to cattle by being chased by dogs. Every man knows that a drive of a couple of miles or so and an hour's worry with a dog is what no herd of cattle can get over in a week's time; besides, there is still another economic side to this matter. I have demonstrated in more than one instance the now-believed doctrine that the farm dog will transmit hog cholera, carrying it from the diseased droves at home that he has been accustomed to be with to the healthy drove belonging to the neighbor. I could give names and actual instances of farmers in my own neighborhood, more than one of whom have lost their hogs quite shortly after the visit of the dog with his master from the farm yard where the disease was in existence. For thirteen years of actual farming, where for years I had hogs by the hundred, up to as high as a thousand at a time, I never had a case of hog cholera among them at all, and I believe it was owing to the fact that I publicly advertised in our county papers that I would prosecute any man who brought a dog onto my place, or who drove a load of hogs that had died of cholera along the road in front of my premises, provided disease should follow this occurrence among my own hogs. I would not knowingly kill a neighbor's dog unless found on my premises, but we never hesitate to remorselessly shoot anything in the shape of a dog who invades our premises or travels that way on the road. A dog has no rights that any man is bound to respect; nor am I alone in this view. I have Scripture authority for my position. My readers will recollect the passage, "for without are dogs and sorcerers," etc. As I stated at the outset, I shall seriously impinge upon some of your prejudices; but if I set you thinking of this subject I am sure you will in time agree with this statement. The dog is a nuisance so far as cattle are concerned; he is a nuisance so far as the children are concerned; he is a nuisance in the house; he is a nuisance outside, breeding fleas, bringing pestilence, and bearing in his presence nothing but what is obnoxious to any cleanly and well-disposed person. Of course you are attached to him, and you stand in substantially the same position as the persons described by Pope, who "first pity, then endure, then embrace." I challenge my readers upon a fair consideration of the matter to deny the truthfulness of that conclusion.

I have thought that one or two instances which I have known of a personal character might not be amiss. The picture of the little girl that is given is that of an actual occurrence. An innocent little girl of seven or eight years of age puts out her hand to pat the dog on the head as she had done many times before. The moment was inopportune; for some reason, and no one will ever know why, this worthless cur, petted and pampered at home, saw fit to snap at and bite the child. It was the merest scratch of a tooth upon the cheek, but almost while I write that child is passing away, in the arms of her grief-stricken mother, with all the horrors of hydrophobia. Nor is the other cut less truthful; a friend in the city of Chicago, Mr. Goodwillie by name, stooped down a moment one day at home, after dinner, to pet a half-grown puppy, when the animal without any previous symptoms of hydrophobia (nor, indeed, will it ever be known that the dog had that disease, for he was killed immediately after the occurrence which I am about to relate), as he was in the act of petting, the worthless cur seized him by the hand and hung there with such tenacity that Mr. G. was compelled to choke him off with the other hand. In less than three months' time Mr. G. was taken with all the symptoms of hydrophobia, and he was so cognizant of the attack when the spell came on that he called his brothers to hold him, and warned his mother and sisters to keep away, saying, "I don't want to bite you, but I know I shall have to if you don't keep away." My readers can picture to themselves the agony that filled that family during the succeeding few hours after the spasms came to be certainly known as hydrophobia until death came to his relief. I maintain that all the dogs in this Republic are not worth two such lives as those just named, actual sacrifices to worthless curs. I could fill pages in recounting instances of the character and kind just named. Only two or three weeks since a family dog, on a farm near a village in New Jersey, an animal considered by his master to be very valuable, suddenly seized his child while in the act of playing with it, biting the child in such a horrible manner that it will be forever disfigured, and then attacking the mother who attempted to save her child, and biting her in such a way that, whether mad or not, it is doubtful if she recovers, and, then, as if to fill the chapter of horrors, when the father, attracted by the screams of wife and child, put in an appearance and attempted to brain the brute, he was himself attacked and bitten so as to be horribly disfigured. To say nothing of the terrible wounds inflicted, here is a father, a mother and a child who must forever suffer untold mental anxiety, lest they, too, come to death by the dread disease hydrophobia. I hope, dear reader, after reading this chapter you will take immediate occasion to feed your dog about two grains of strychnine or put a chunk of cold lead through his head, and prevent the possibility of occurrences similar to the above in your family. It is a well attested fact that there is something in the organization of the dog that renders him liable to a mad attack in either extreme cold or extreme hot weather, and the trouble is you never can tell when the thing is going to happen. There is but one safe way and that is (and it ought to be the motto of every farmer)

KILL EVERY DOG ON SIGHT.

I have tried the best I know to conjure up some excuse for keeping a dog. They are a nuisance to the cattle; they are a nuisance to the hogs; they are a nuisance to the sheep; they are a miserable nuisance in the house, and they are as often a nuisance outside as otherwise; and I can conceive of only one place where a dog is needed, or where a man would be justified in keeping one. In large cities, where there is a liability at night of the premises being burglarized, I can there conceive of cases in which a dog might serve a good purpose. I don't know that I ever read or heard of any—but I can conceive of such a case; but it seems to me that in this case the dog should be kept much as we keep a lion or a tiger, chained, or in a separate apartment or yard, fenced in by himself, to be let into the room to be guarded only at night, and removed to his pen in the morning, never having a chance to reach people outside. It would be interesting reading, perhaps, to go into a detailed statement or calculation of what dogs cost the people of these United States. I can make a guess. I would place their number at 10,000,000. Perhaps that is too high; call it 5,000,000. My readers can figure the economic side of the question to suit themselves. If each dog actually consumes in a year twenty dollars worth of food, the price of one good hog, of five sheep, or a cow, they can understand without much trouble how much we lose by reason of the curs—the worthless good-for-nothing curs—in this land of ours. Dogs, like the horns, are a nuisance, and "dogs ought to go," and the refrain ought to echo all across the continent. While penning these lines there comes over the wires from Alabama this dispatch: "A few months since a dog, supposed to be rabid, bit seven negroes at a public meeting they were holding in one of the interior villages before he could be killed. To-day one colored man died, and all the rest that were bitten have given up in despair and taken to their beds, and all expect to die." Poor things! And yet there are men who will allow themselves and their wives and children to be continually exposed to the possibility of such a death right at home.

Say, neighbor farmer, I wish to propound a conundrum to you: "Can you tell me why it is that a purp can't see until it is nine days old? Give it up? It is to give you nine long, straight, thinking spells, and nine good chances to kill it. What a wise provision of nature! How different the case of the calf and colt. Less hardy by nature, they will yet when just born get up and flee in ten minutes with the dam for life, but the purp can't even see its way out, much less escape. "Never thought of it in that light before?" Kill it at once, and so spare the human race possible misery and death.

CASTRATION.

Since writing the article on Castration, the author has had much experience on his own herd, and the benefit of the experience of many others who have adopted his mode of castration, and he is more than ever satisfied that it is the way to castrate bulls. He has, therefore, added three cuts showing the position of the parts and the manner of operating. His plan understood on older bulls will need no further elaboration in castrating calves. Let us suppose, then, the bull is safely inclosed in one of Haaff's portable chutes, with a bar behind so that he cannot back out or kick during the operation. The bar should be so low down that the operator can readily handle the parts.

Seize the scrotum or bag with the left hand, not grabbing around the testicles, but taking hold of the skin or sack of the testicles from behind, and turning them around to the right, so that the front part of the testicles are turned towards the operator. This refers to the position of the scrotum or bag. Now take a sharp penknife or small-bladed knife in the right hand, and holding the thumb only upon the blade, say an inch back from the point, grasp it firmly, and drive it into each testicle way up at the top, and with one continued slashing, driving cut bring it down to the bottom of the testicle. Do the same way with the other testicle.

Now, if you release your hold and examine a moment you will see that the two testes have dropped out of the skin, and are hanging by the cord and by the connection which they have with the sack, and you will see on the testes that you have made two well-defined longitudinal cuts into the soft part of each testicle. Don't allow the testicles to hang down weighting the animal's cords, but still holding with your left hand, pinch out the soft part of each testicle between the thumb and ends of the fingers of the right hand. Pinch and squeeze only the soft part while you hold with the left on to the sack. When you have removed all you can by the pinching process, by looking along down the cord you will see the epididimus. Taking the point of the knife make a transverse cut horizontally across the cord and epididymis. This will destroy the connection, and prevent the bull from being "proud," as it is called. Restore the cord and parts to the scrotum, gathering the cut together, and removing nothing more from the parts, unless, in case of animals abnormally large, a small part of the internal skin may protrude from below in which case trim it off with a knife.

Now, in the case of the calf, lay him down, and while an attendant holds one hind leg well forward, and rests his weight on the neck and shoulders of the calf, or between the legs, so that he cannot move, you will cut in just the same way as in the case of the older bull. You will do the same pinching, and you will make the same little cross-cut on the epididymis carefully, because the parts are tender, and because if your knife is very sharp there is a possibility of your striking through and cutting your own fingers. Restore the parts of the calf to the inside of the scrotum or sac, and let him go. Of course you have done the same thing with the big bull; that is all there is of it. There is a possibility that the parts of the scrotum may unite and heal by "first intention." In case there seems to be much swelling, run the bull into the chute; examine him carefully, and see that there is a proper vent at the bottom of the wound. Now, if you made your first cut right, as it should have been made, there will be no healing by "first intention," for the bottom of the scrotum will be kept open by the hemorrhage and pus that naturally follow.

Now, what are the advantages of this mode of castration? First, your steer will always have a "Cod." This is a sign of a fat steer among the buyers, or at least, supposed to be one sign. Secondly, in case of a two-year-old bull, if you have also dehorned him, which, by the way, may be done at the same time, you will have after one year a straight, perfect appearing steer, and no one will ever guess that he was a bull. I know one man—yes, I know two or three of them—who make it a business to buy up yearling and two-year-old bulls, and dehorn and castrate them after my process, and sell them as straight, regular steers the year following. And they are straight, regular steers, and there should be no discount on them whatever. So much for the combined operations of dehorning and castration. In the third place, the advantage of this mode of castration is that it prevents the possibility of internal hemorrhage.

I suppose that not one farmer in over fifty knows how to properly castrate an old bull. It is astonishing what a credulous, unsophisticated, self-deprecatory style your ordinary farmer will adopt. He is afraid to do anything that is out of the ordinary beaten track. Now bulls—old bulls, that is—that die from the operation of castration, do so by reason of internal hemorrhage. You can see in a moment that to take these testicles, after the scrotum was cut, as they hang down there, and draw them down as far as possible, and then sever them with a knife, or break them off, and so let the cord and muscle contract and spring back above the opening or neck of the scrotum, lying up there in the body of the animal and bleeding, the blood trickling down to the neck of the scrotum and coagulating at that narrow point or orifice, is simply to kill the animal, for if the orifice closes then there is internal bleeding or hemorrhage, and of course pus follows; then blood poisoning and death. Now, my way is not to cut off the cord, and hence as the end of the cord is down in the bag or scrotum, and is held in position by its connection with the scrotum, it is simply impossible that there should be any internal bleeding, and hence I do not see how it would be possible to ever lose a bull by this mode of operating. I know this much: Myself and others have pursued it with success and subsequent satisfaction. The figures show the left hand grasping the scrotum, and in the act of turning it around.

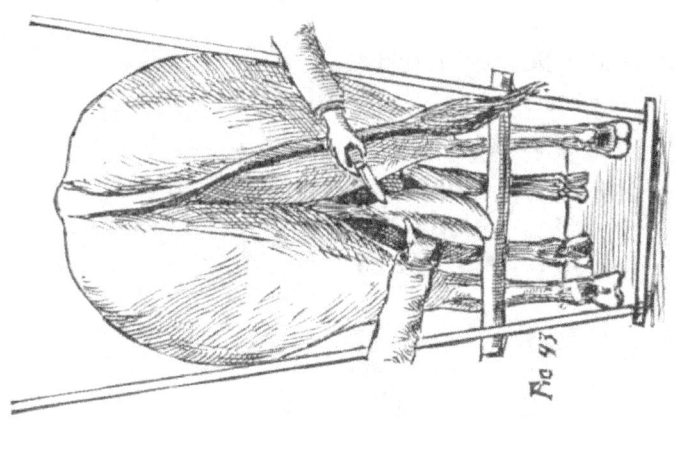

Fig 43

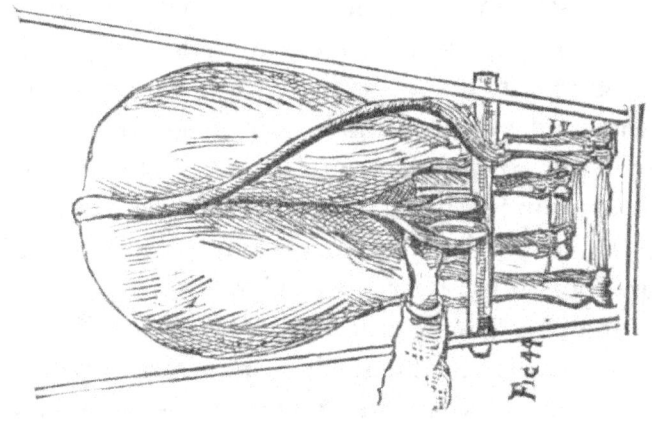

Fig 44

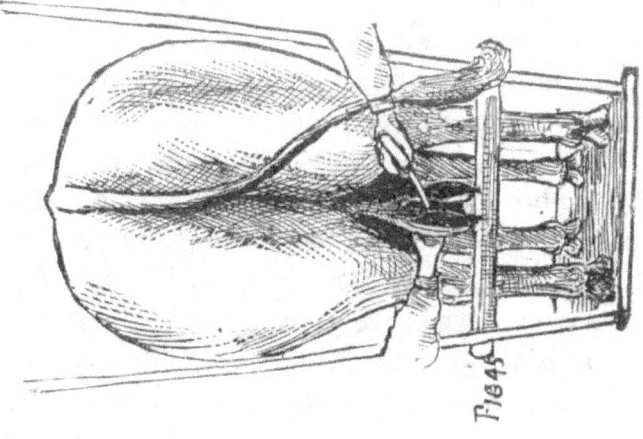

Fig 45

161

Another cut shows the scrotum turned fully around, and the knife point placed at the neck above where the cut is to begin; and I will say right here, cut steadily and firmly clear down through the entire length of the scrotum and testicle to the bottom thereof. The third cut shows the knife placed in position on the epididymis for the cross-cut of the cord and epididymis.

CATTLE TAGS.

It is now about fourteen years since my attention was called for the first time in a practical way to the matter of marking cattle in some way so as to distinguish them one from another. There are many different ways in common use, but, so far, but one effective mode has been found, and that is attended with so much trouble and inconvenience to the owner, and so much hardship, if not of cruelty, to the animal itself, as to be particularly obnoxious to every cattle owner who cares for the appearance and comfort of his brutes. I refer to the practice of branding. If we recollect the definition of cruelty, "the infliction of unnecessary pain," we shall find in connection with the operation of branding that this practice is attended with more suffering than any other save that of castration or spaying. When it is remembered that the nerves are principally on the surface, and that the hide or skin of the animal is simply so crowded with them that it is impossible to place the point of a cambric needle upon the cuticle without touching a nerve, it will be at once seen that to place a red hot iron upon the animal's hide, burning the hair, and through the cuticle into the corium or true skin, covered by so large a surface as that occupied by the branding iron, must necessarily destroy very many of the nerves, and must be particularly painful by reason of partially destroying many others adjacent to the place where the iron touches; but to attempt to brand with an iron not red hot is doubly painful, and produces the most excruciating agony that the animal can experience. I have handled many thousands of cattle on my ranch in Henry County, Ill., yearly, for more than ten years last past. I have declared time and again that I never would brand another animal, and yet, while I have not personally applied the iron, I have been compelled to know that the iron was used, because men bringing cattle to my pastures necessarily insisted on that mode as the only way of being able to certainly distinguish their own cattle from others. No man living can drive 50 or 100 cattle to a pasture remote from his own, and after some months have elapsed again distinguish his cattle. I have seen this thing tried over and over again, and I have had it tried, and trusted to it to my cost, by men who were old cattle men, and were certain that they were an exception to the general rule. There is but one way to handle cattle if it is desired to distinguish them from other people's cattle or among themselves. The animals must be branded or marked in some way so as to be able to be positively identified on any occasion. Some men mark the ear, slitting with a knife or sometimes chipping out a piece, or more frequently cutting a hole in it with a wad cutter or other instrument. A dozen years since I was led to believe that Dana's cattle tags would fill a long-felt want, but they are little better than nothing, as I have found in a thorough trial. I have used many hundreds of them myself on my own cattle, and I have depended upon them as a distinguishing mark in the case of very many other herds of cattle, and the result has always been uncertainty, dissatisfaction, and frequent trouble; so that instead of being able to distinguish the cattle by these tags we were compelled to resort to any other way that ingenuity could devise. The trouble with all these tags is many fold. In the first place they are so small as to be more or less obscured by the hair of the ear, which renders it well nigh impossible to tell if the animal has a tag or not at the distance of five to ten rods. In the next place the ear and the head itself are almost perpetually on the move, and between the various movements of the head and ear no man living can do more than guess at what these tags read. In the third place the cattle tear these tags out of the ear by rubbing against any obstruction like the head of a nail or a projecting end of wire. Whatever the obstruction may be, if long enough to catch on to either the tag or the hole made in the ear to receive the tag, it is almost certain to be torn out, and most frequently the ear is left so badly disfigured as to forever mar the beauty of the animal. Another kind of tag which it was supposed by some would obviate the difficulties met with in the use of Dana's tags is that which is put into the ear by shoving one half into the slot in the other half like a sleeve button; but the trouble with these tags is that while they are not liable to the objection of being torn out, they are liable to, and do in thousands of cases, rot out, because by the compression of the parts of the tag the circulation of the blood in the ears is retarded, or else if the tag do not clasp the ear so strongly as to stop the circulation, it will so impede it that extreme cold weather will freeze the ear, and succeeding warm weather will cause the part to decay and the part will drop out. Besides, this kind of tag as well as the Dana tag are both liable to a fourth objection, namely, they are too small to enable one to distinguish the cattle at any distance. The old style of marking cattle with hog rings, or with leather tags, or tin, or other metal tags sustained by hog rings or by wire rings in the ear, are all alike open to the objection named, they are almost certain to be lost by the animal scratching against any projecting substance whatever as a sliver, or a nail, or wire. Another mode of marking cattle resorted to by some men is that of cutting the brisket or dewlap. This plan of marking cattle is chiefly objectionable from the fact that of itself it constitutes no mark. For some years my dehorned cattle were as thorougly marked as it was possible for cattle to be, but when after observing my success, my neighbors themselves began to adopt the practice of dehorning and use it on their own herds, it was no longer possible to distinguish my cattle by that way of marking; so, too, to cut the dewlap simply amounts to no mark at all.

This, however, is a fact that is well established. It is a perfectly safe practice to cut the dewlap or brisket; that is, I mean to say it is not attended with any danger to the animal, nor do any serious consequences follow it. I have given this subject of branding and marking and tagging cattle much thought, and after very slow and careful deliberation, I have come to the conclusion that I have discovered the only true way to mark cattle. The illustration herewith annexed shows my new patent mode of marking cattle. As will be seen by inspection of the cut my plan is a metal mark or tag attached to the brisket or dewlap of the animal. These tags are brass nickel plated. Let us now consider the objections that have been raised to all the other forms of marking cattle and see whether they apply to this. First: The suffering of the animal. To punch two or three quarter-inch holes through the brisket of the animal at the proper place, as my instructions show how to do, and which always accompany the tag, can do the animal no injury, and can produce but momentary pain; so, then, this method is not open to the objection that there is to branding or to slitting the ear. Second: Liability to be torn out. There is no danger in the case of my brisket tags of their being torn from the cattle. In the first place there are no openings through which a wire end, or a projecting nail, or a splinter of wood could be thrust, and, besides, the animal is not so much accustomed to rubbing the brisket or dewlap at the point where I place the tag, and, further, if it does rub it there is no danger of his displacing it. Third: The objection of not being able to distinguish the animal at any distance by means of the tag. First, not true of my brisket tag. This tag is so large, being a piece of metal, pure silver in color, showing a surface at each side of the brisket of two inches in width by from three to six inches in length that from whichever side you look at the animal this tag is always in plain view. But the best thing about this tag remains to be told. Unlike all other tags it has no letters or marks painted on it, but the letters on one side and the figures on the other are never less than an inch in length, and are always made by stamping them out of the metal. This is so that when placed in position on the animal the color of the hair or hide as a background is what forms the letter. A moment's inspection of the cut will show what I mean. This tag may if desired be put well forward into the animal's flank; and I believe that a simple flank tag, with the ends properly rounded, is the true way to mark a horse.

The directions for the use of Haaff's Tags will explain all that is necessary to be known. The holes are first punched through the brisket or flank at the point desired, which is given in the directions; the animal should then be turned loose and allowed to run from two to four weeks, or until the wounds made by the punching of these holes are thoroughly healed. It is then but a moment's work to insert the tags, and the vexing question of how to distinguish your cattle is forever settled. You can mark and number every animal from one up to one hundred or one thousand; as many cattle as you have, you can keep the record of each animal in any book that you please, and the plainest, cheapest kind of a book is all that is necessary. Suppose that you have two hundred head of cattle—beginning with your oldest cattle you mark them down; first cows, then steers, just as you please, and you come to know what number 1 or number 10 or number 47 refers to, and should an animal be lost, any one seeing that animal could not fail to distinguish it from every animal living; should an animal be stolen and the tag be removed you still have positive marks of identification, because these two or three holes which are cut in the brisket, and which are allowed to heal up, can never be successfully imitated. Any man can steal a brute with one of those ear-marks and easily counterfeit the hole left in the ear, or destroy it by tearing it out. Any man can rebrand an animal by first burning with a hot iron so as to obliterate the brand and then of course by applying his own brand; but no cattle-thief can punch holes in the brisket so as to leave a mark which will give the same outward appearance after the

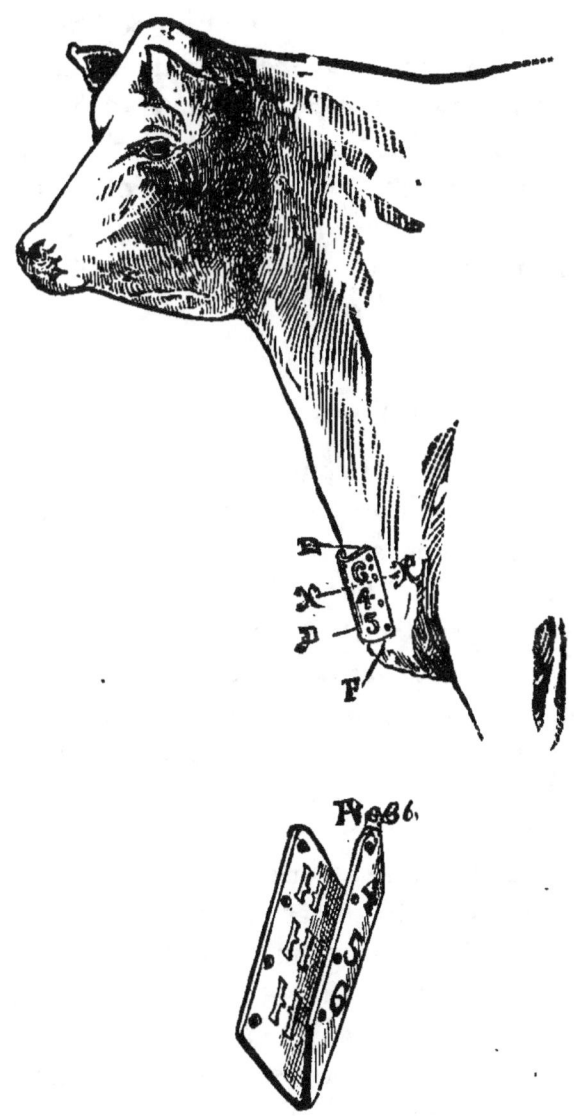

Haaff Brisket Tags have been removed. The wearing of so large a tag on this point on the animal makes of itself a mark that no living man can counterfeit. I need not tell my readers what a saving is effected by the use of a good tag. The loss on the hides of cattle that have been branded varies in different localities, but nowhere is it less than 10 per cent, and if the hide of your animal is worth in the market $5, you will save just 50 cents less the few cents cost of purchasing my tag. I believe that my tag will be as universally adopted in the near future as has been my practice of dehorning.

SOME REFLECTIONS AND COMMENTS.

Paul, you know, says: "Let not him boast himself that putteth on his armor, but rather let him boast that putteth it off." I have not intended to boast. It is true the personal pronoun frequently occurs in these pages, but it is every time the experience of your friend and well-wisher who has put on the armor and worn it and never taken it off until assured of having won a thorough and complete victory over all objection that could be raised. In offering the public my practical method of dehorning cattle, I did not boldly rush to the front until by some years of experience I had ascertained that I had a sure and certain and a perfect method of removing the horns from cattle; and I boldly assert that not a single case can be found in all the thousands and tens of thousands of cattle that have been dehorned during the past year, since I made this matter public, that is in any sense a failure, and wherever bad results have followed it has been due to carelessness in the method of operating, or in the use of tools different from my own. I have never complained if men saw fit to use stiff-backed saws or meat saws; I have simply had to wait and they themselves would see in a very short time that these tools were not proper tools to use in dehorning cattle; so, too, with my gouge; I didn't object if men chose to use a knife upon the calf's head, or hedge trimming or other kind of shears; I simply said that plan will not work; I said the same thing about burning the calf's horns with a hot iron. I have letters from hundreds of those men in which they admit what I have insisted and still insist is the way to pursue, that to dehorn cattle and do it properly it is necessary to use proper tools made for the purpose. I now insist that to mark cattle and mark them properly it is necessary to use the proper kind of a mark and properly place it upon the animal in the proper way. I have been urged by scores of men for more than a year past to give this cattle tag to the public. I have refused to do it until now, assured of its positive and certain success in my own mind; so also in the matter of an apparatus for heating water. My readers will bear me out in saying that I have urged upon them the advisability of adopting a simple and cheap method of heating water. We have embodied that idea in my new tank heater, and I simply rise to ask this question: What is the use of a farmer paying from $18 to $25 for a water-heater when he can get a better one for half the money? I give Haaff's Tank Heater to the public after much deliberation, believing that no farmer will use it and not say that it is a right thing in the right place. It is surprising how far behind the age is in developing any marked improvement in agricultural implements as compared with developments in other lines of public usefulness. It is as true to day as it was a hundred years ago that—

> He that by the plow would thrive,
> Himself must either hold or drive.

Two generations have passed in this country, and in that time the railroad has supplanted the stage coach, the mail service of the country has been so thoroughly changed as to bear little semblance and less relation to the so-called mail service of the country two centuries past; the use of electricity for that purpose and for motive power; the electric telegraph; the use of the telephone; the express service of the country; the style of erecting buildings of a fire-proof character; the work in gutta percha, and in branches of art of a similar character too numerous to mention, are all of them the products of the last fifty years; but it is a fact that little or no improvement has been made in our method of turning over the soil; the plow of fifty years ago was drawn by horse, and the plow of to-day is drawn by horse, and I am not prepared to say that the labor has been much lessened, and as to the advent of the sulky plow I question the character of the improvement. Sulky plows are notorious horse killers. There has been little or no improvement in these lines of farm work, and in our methods of securing hay and grain there is but little change for the better. I believe that the old-fashioned hay sweep that was used when I was a boy will gather in hay every whit as fast as your modern so-called hay carriers and wide sweep rakes. It is true we have the grain self-binder; this, to my mind, is the one single and signal exception to the rule of no progress in the art of farming. I know I shall be laughed at, and shall probably be mercilessly scored for uttering these sentiments, but I believe them to be true. There is something singular in thinking over this matter, what does it mean? how is it that we have never had a successful farm engine? There has been absolutely little or no improvement in the matter of motive power on the farm, either for threshing grain or for pumping water, for sawing wood, for cutting feed, for making hay, for plowing land, in fact for any kind of farm work save in the single one thing of the use of the self-binder. I hope, and I believe, that in some matters it may be agreed on all hands that I have begun the march of improvement, and if so I shall feel that the best thirteen years of my life, devoted exclusively to the farm, were not thrown away. I believe it is possible to give my countrymen a grass that shall withstand the severest drought we have ever known, and yet yield on any ordinary soil two tons of delicious hay per acre. I believe it possible to produce such a grass that shall bear a leaf or haulm more than double that of timothy grass, and that shall produce at the same time a weight in grain that shall be heavier than that afforded by Hungarian grass, and a grass that shall be perennial, flourishing equally well in sandy and in black, heavy clay soil. It is my purpose to strive to produce such a grass, and teach my brother farmers that it is perfectly practicable for use on the farm. I believe it is possible to produce a farm engine that shall be adapted to all the purposes to which an engine can be put on the farm, that shall entirely do away with the windmill for pumping purposes (a machine that is usually at a stand-still when water is most needed on the farm). I believe that such an engine can be produced to cost not more than five cents an hour to run it, in which the danger of explosion shall be nothing, and in whose use the danger of fire shall be next to nothing. I believe that within the next ten years, and I hope in less than five, I shall be able to produce such an engine and offer it to my brother farmers, and be able to say of it as I do of dehorning, and of my tags and water heater, "It leaves nothing more to be desired." Until then, adieu!

THE END.

APPENDIX

This book has been delayed a straight month. There is no use in criminating any one: to blame the printer is useless, and I don't take to it myself very kindly. Perhaps the cuts here given may in part recompense you, dear reader, but if not, "go for the horns," and your temper will sweetly wear itself away.

I believe the new Webster Chute is a success. If you choose to build it, write to him at Marysville, Kansas, and get his circular, for it is a patented chute—or at least a patent is applied for.

At the last moment I add two cuts showing a big improvement on the use of the "Jewel," and which I shall patent for those who buy my book and tools. This will do away with the "Bull Leader and Pulleys" entirely, and I am glad of it, although I went to large expense with a manufacturing company for a new cast of farm pulleys; and you that have them will find them like my saw, useful in many ways besides in dehorning. Note in the cuts that I make the Jewel rigid, and put two straps of iron (or you can use strap hinges) all along the back side, and bolt it permanently to the plank. I also do away with the log chain and use a double piece of rope instead. This can be slipped through a hole in the plank or a ring, and hauled up tight to the neck by a handspike as we did the chain, or it can be slipped through a hole in a small round stick which can be used on the back side of the plank as a windlass. The same is true of the rope shown over the nose. You can pry out on it with a handspike, or use a small stick as a windlass—say six inches long—and turn it up tight to hold the nose. With the nose

APPENDIX. 173

and neck held, dehorning will be made simple with any chute. Of course now the plank must be run out two feet instead of one beyond the chute and tapered down a little if found best, and the nose may be tightened by a wedge or wedges as well as a handspike or a windlass. Write me, anyone who has trouble, or wishes to make inquiry, for I am yours to serve. I add Mr. Butz' letter, received at the last moment, and I add in final conclusion, "The truth is mighty and will prevail." Go thou, reader, and do likewise.

I add a letter from Mr. Arnold, of Blue Hill, Nebraska; it is good reading. He reports, March 28, having dehorned 1305 head of cattle, and adds, "I don't know it all yet." And now let such scrub editors as he of the *Montana Live-stock Journal* growl, and tell his readers—as I am told he does—that any old saw will do the work just as well as another.

Compare the idiot who never took off a horn with this plain Brother Arnold, who lives on his own little place, and has tried it, and "don't know it all yet."

I hope Brother A. won't use a knife any more; otherwise his plans show good sense, but the knife business I discarded after more trials than I care to tell here. Use the outcutter first on calves, and then the gouge, and be sure to grind it on the back so as to keep the jaws slanted down, and not on a straight horizontal line. The idea with the gouge is to lift and pry up and out, as well as to cut, and I wish I had made the gouge figure to look more that way.

I add to the Index which is hereto attached a separate list of names of papers and of "some dehorners" among the thousands I have received.

I should be derelict to my duty as a man, and wanting in proper respect to my friends if I did not in this volume renewedly mention my deep sense of personal gratitude to the "Forty Farmers" who turned out in the January blizzard of 1886 to help defend me personally, and the cause of dehorning cattle, against the assaults of the so-called Humane Society. The names of Taber and Gilbert, of Jennings, and Heaps, of

Powell, and Arnett, and of our distinguished attorney, Dunham, and those others whose names would largely swell the list, are ever held by the author in kindest remembrance.

H. H. HAAFF.

Box 193, Chicago, Ills.

Blue hill Neb march 28 1888

Mr Haaff

i reply to youre leter about youre book i want it i have youre gog and saw i got a year ago but calves is the hardest thing i can dehorn it seames to hert them worse i have dehorned 1305 head since last octtober and about that maney before then well i have learnt lots about it and dont think i no it all yet when I dehorned the first Five head I new it all and i kill a calf by trying to dehorn it with my nife i let the nife slip and it run to its brain and it dide i dehorned a big bool and smothered him to death then i donup a nother bull and the man said it bled to death he got out and fit a half day with a nother bool then i went to dehorn for a man that had 50 head and one por yearlin bled to death i cut him open and there wasent blood a nuf in him to stain my hand these men gave me H but i went ahed bond to learnit so now i go far and near i maid me a rack that i can hold aney Bull or wild texses stear it is long a nuf to let there sholders in then they stand on a Plat form to keep it from falling over i bilt five rackes be fore i got the thing rite well now bout the horn i take the horn down in deap so as to take the making of the shell horn off and when they get well they are smoth muleyes i go to the senter of the crease from the horn to where the scull raises and the first that i don evry one has got a stub groan out i did not go deap a nuf to get the making of the shell horn

I run my nife round cut clear to the scull then take my saw and take it off i can take the hornes off very quick when they blead to much i ketch them and put on some puff ball and if that dont stop it i take a neadal and run threw the veain and ty a thred around and stop it that way i tride to sear it i dont like that way i find by runing my nife in a strait down forming a dish in the skin to hold the blud that it stopes itself much quicker than when the veain is cut squair of send your Bok that treeats on dehorning

hoping to hear from you By return mail

John Arnold Blue hill

Webster conty Neb Box 72

i had Mccleary to send for youre tooles for me he lived in hastings he is my Brotherinlaw that is the way i got youre Bok saw and gog

I see that the printer has set up Brother Arnold's letter *verbatim et literatim*. Well, no matter! Let it go so. No offence is intended, and now perhaps the "Pharisees" will believe for once that I have the original letter.

APPENDIX.

Hope, Ill., April 16, 1888.

Mr. H. H. Haaff.

Dear Sir:—I met you in Danville, Ill., some time ago, and thought you were a crank, and sure enough you have twisted me completely over. If you are ever in our county again would be glad to know when, as several of our largest stock raisers are coming over. I am willing to have my herd sacrificed first. Please send circular.   Yours truly,

J. K. Butz.

Good for an honest farmer.

PLAIN DIRECTIONS FOR DEHORNING CATTLE.

CALVES.

Use the outcutter first. Hold the head still and turn the cutter until you feel that it has cut through hair and hide and membrane and into the bone. Keep the cutter sharp. Keep the calf cool and don't dehorn him until to-morrow if he has been running. Have the gouge sharp and well tapered at the cutting edges so that it will get down into and under the embryo horn and lift it out. Use flour if there is much blood.

OLD CATTLE.

Have your chute ready, and plenty of good help. Kill off the dogs, and don't drive the cattle much the same day you dehorn them. All these things excite them, and excitement is bad. Have your chute so built that you can secure your cow *the first time she runs in*, by a pole or a bar behind her. Put the best man at this post, for there is more bother about not shoving in the bar behind than about all the other bars. The steer thinks he is going right through first time. If he backs out he wont be so ready to come up again, and you will begin yelling and punching, and that is very bad. Secure your bull first time trying; then on with the Jewel and nose-rope and off with the horns, and "let 'er go." Where the horn lies well down into the head, make two cuts, one on top first, starting on a big steer from a quarter to a half inch back into the hide (that is, into the matrix), carry the saw well along at same proportion of hide and hair until you pass the middle of the horn. Now reverse the saw and cut from the under side to the point you stopped at. Open chute and let 'er go. Watch for cases of bleeding; use puff ball or flour, and of course, if needed, slip a pin through the artery and wind a thread around, and use puff ball or flour. Keep the animal quiet and don't be scared. If you get a good ready and do your own dehorning without having a campmeeting, you won't have any trouble or loss. It sometimes happens that a frozen horn or a diseased head will bleed badly. Watch them, and do as I tell you above. It's very strange that no one ever has any trouble where I dehorn. The secret lies in the fact that I *will have quiet and will not use a meat or a carpenter's saw*, which are sure to go wrong sometimes and cause loss. You can see what men in this volume say, and judge for yourself. I am of opinion that my chute, with a ten inch plank on the bottom, and bars in front so to slide as never to drop down, and a bar or bars under the belly and one behind (see article "Chute"), and the new arrangement of the Jewel and nose rope leaves nothing more to be desired about dehorning older cattle.

H. H.

WEBSTER'S NEW CHUTE.

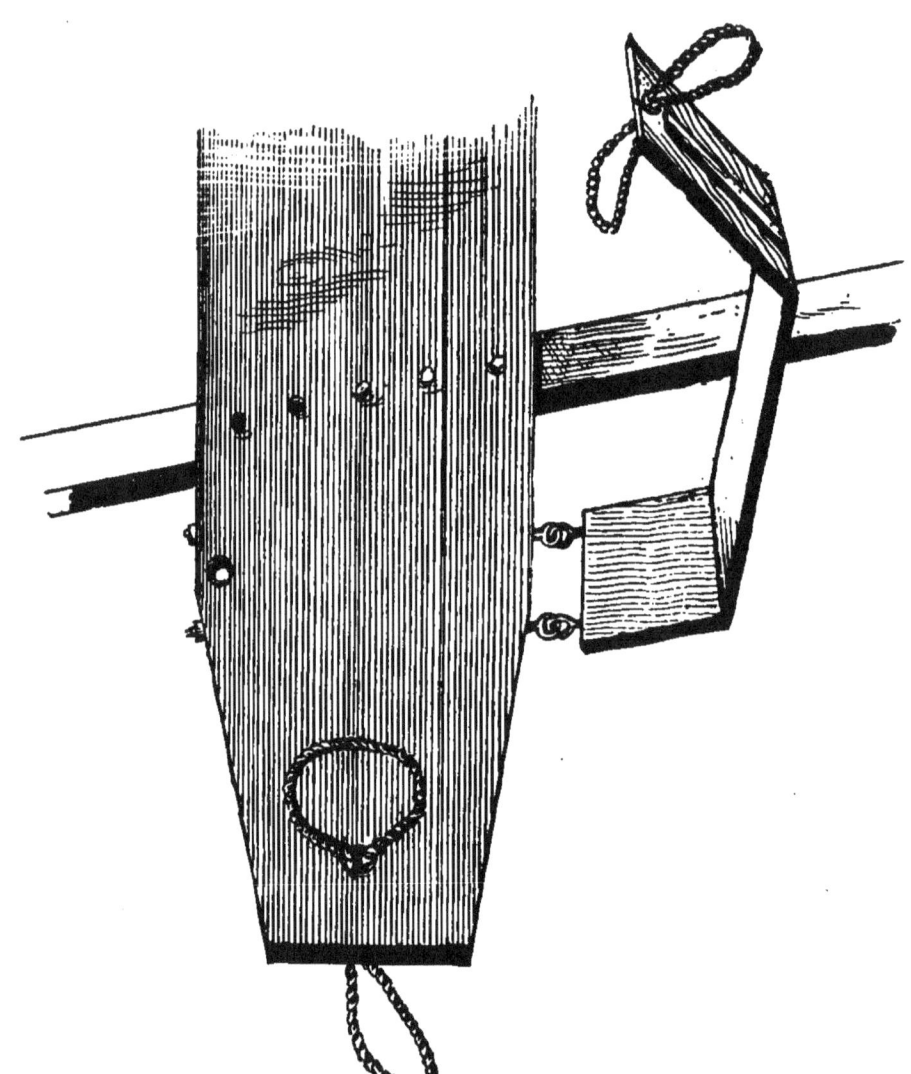

THE JEWEL AND SIDE PLANK.

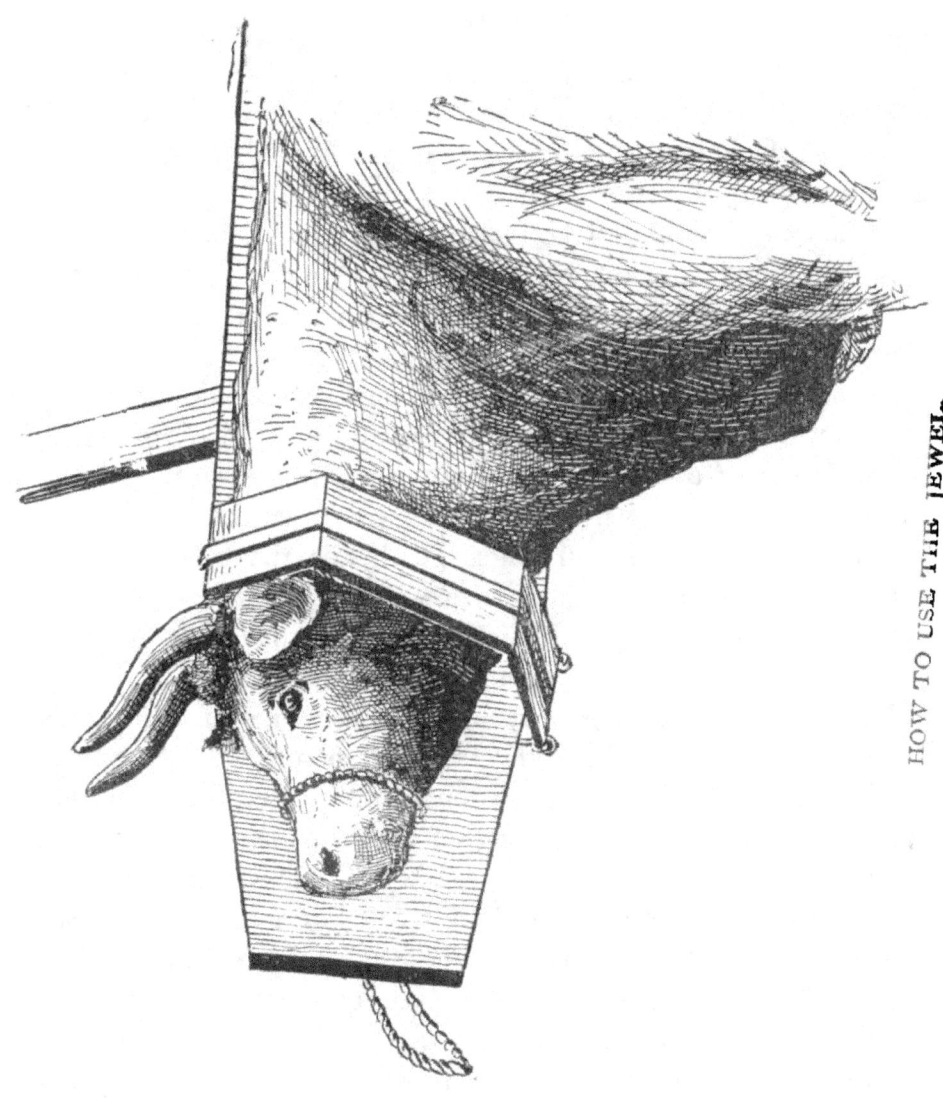

HOW TO USE THE JEWEL.

HAAFF'S BRISKET CATTLE TAGS.

(Patent applied for.)

Like every new thing, you must get used to this Tag, and when you do learn how to use it no man can hire you to brand your cattle Send 10 cents for a sample Tag. You will never regret trying this Tag

This Tag is brass, nickel plated. This Tag may be secured to the animal's brisket, either as soon as the holes are punched, or after the holes have healed up, (which I think is the better way,) same as with any other kind of a tag.

Punch two quarter-inch holes in the brisket as high up as the brisket will allow, so as to correspond with the Tag when doubled up, (for these Tags are made of many sizes,) slip in the coupling tins, being careful not to crowd the Tag down too much onto the skin; then put inside the tins a copper rivet ½ to ¾ inch long. Do not rivet too tight. It is well to use a three cornered file in the holes if they are too small, so that the tins will not turn. No danger of being torn out. No danger of freezing or rotting out. Can be seen many rods away. Gives both the initials of your name and the number of the animal. Saves you 50 cents to $1.50 loss on each hide by branding. No disfiguring of ear or hide. Is a positive ornament to the animal, and costs only ten cents each by wholesale in quantities; and, best of all, no cattle thief can ever counterfeit this mark, and if he takes it off, he inevitably leaves the place in such a worn condition that it is an unmistakable mark forever. It is the only Cattle Tag worth the name. These Tags are ten cents by wholesale, and agents should charge fifteen cents each put onto the cattle.

Send one dollar additional for punch and nippers, etc. After you have put on the washer, cut the copper rivet off, and rivet the end a little with small hammer. One crease with the file is enough to let in the tins.

INDEX.

SOME DEHORNERS.

Ashley	85	Holm	95
Adams	97	Hoard	68
Arnold	94	Kansas College	118
Arnold—(Appendix)	174	Kull	101
Askey	101	Kelso	96
Brown of Amy	70	Luce	84
Bishop	80	Lindley	86
Berry	99	Miller	94
Bond	120	Morris	80
Bauer	102	Moses	109
B. F. R.	116, 123	Moore	126
Butz—(Appendix)	175	Secretary Newton	66
Chandler	60	Pierce	81
Carter	80	Peters	84
Constance	80	Richardson	102
Cox	83, 96	Richards	38
Dr. Cutts	106	Roberts	123
Campbell	96	Subscriber	102
Crane	98	Schreiber	114
Ed. Cheever	62	Sawyer	80
Clark	122	Stevenson	82
Davis	86	Stoops	85
Col. Davidson	87	Tillottson	95
Davisson	93, 127	Tebow	83
Erb	81	Underwood	98
Frisbie	86	Webster	34-7, 180
Frye	86	Whiting	41, 89
Fletcher	88	Wood	60
French	97	Way	129
Foote	100	Walton	86
George	60	Weber	85
Goodwin	60	Waite	109
Gibbs	60	West	118
Gardner	75	Williams	94
Gillus	84	Welch	94
Prof. Henry	62, 4, 5, 123	Warner	97
Hermance	82	Wilson	98
Heath	113	Wattles	100
Hillman	115	Young	84

THE PRESS.

Breeder's Gazette	60, 120	Massachusetts Ploughman	62, 76
Dairy World	63	McHenry Sentinel	46
Farm and Home	62	New England Homestead	95
Farm, Stock and Home	104	Orange County Farmer	62
Farmer's Review	60	Other Papers	63
Farm and Fireside	62	Rural New Yorker	62
Hoard's Dairyman	62, 120	Western Rural	60, 111
Jersey Bulletin	62	Western Resources	82
Live-Stock Indicator	63, 85, 90, 117	Short Horn Journal	118

INDEX.

Agricultural Papers	15
"A Dollar a Horn"	84
"A Screamer"	121
Burning	129
Butt	111
Butz	200
Branding	14
Bone Horn	22
Blow on the Horn	28, 106
Bruises	30
Bulls	48, 68, 158
Bull Leader	49
Bars	50
Broken	52
Bone Horn	10, 52, 55, 6, 7, 8, 9
Brain	54
Boards of Agriculture	63-4
"Butter Potent Bulls"	68
Butter Test	70-1
Bleeding	127, 129
Calves	17
Chute	39, 40-6, 51, 91, 200
Cuts	5, 6, 7, 8, 9, 25, 35, 36, 37, 39, 44, 45, 55, 132, 156, 162, 200
Choking	51
Chain	49
Casting	50
Circulation of Blood in Wound	10, 71
Cud	72
Cattle Tags	72, 91
Directions—(Appendix)	180
Cruelty	73, 75, 91, 111, 139
Cutts, Dr	106
Comments	137
Castration	158
Deer Horns	11
Dehorning—How Invented	13, 16
Dogs	110
Disturbing Causes	16
Discussion of Bone and Shell Horns	23
Dehorning Old Cattle	24
Using a Hot Iron	24
Different Ages	27
Diseased Horns	30, 32
Death from	31
Dairyman's Association	67, 68
Discovery of	13, 91
Dehorn vs. Dishorn	95
Directions	124
Dogs	150, 158
Effect of	71
Engine	170
Frozen Horns	30, 91
Frontal Sinuses	54
Frontal Bone	54
Flour	127
Flies	131
Farm Engine	170
Farming	170
Gouge	19
Guernseys and Jerseys	68
"Gentle Jerseys"	75
Horns	9
A Nuisance	77, 106
Make up	10
Circulation of Blood in	11
Three Horned Steer	11
Growth of	17
Hot Iron	24, 136
How to Build Chute	46
How to Hold Head	49
Healing	53
"Hollow Horn"	91
Heaters	194
Injuries in	33
Jewel	48
Jerseys and Guernseys	68
Knocked off	91
Livingstone	29
The Loop	32
Letters—Hoard	69, 79
Times	88
Losses by	78
Matrix	29, 53, 130
Methods of Securing	32
Mulleys	50, 77
Meeting at Madison, Wis	60
Milk Test	70, 87
Maggots	131
New Method of	43, 180
Nerves	106
Outcutter	19
Polled Cattle	12
Process of Healing	23
Periosteum	26
Power of Horn	29
Pirates	43
Price of	46
Plank	49
Pulleys	50
Parietal Wall	54
Pain in	106
Press and Papers (See *ante*)	
Puff Ball	127, 129
"Putting Something on"	136
Resolutions of Wisconsin Convention	65
Rings	74
Rounding-up	138
Shell Horn	22
Shears, etc	24, 111, 136
Saws	24

INDEX.

Skull Bones	26	The Reason Why	15
Structure of Head Bone	28	The Place When, etc.	16
Stub Horns	29, 53	"Ten Thousand Demand It"	86
Sore Heads	30, 530	Tools	125
Suppuration	31	Turpentine	131
Stanchion	33, 35, 36, 37	Tanks	146
"Sawing off Horns"	46	Verses	74, 110
Shaping the Head	54	"V.S."	75, 79
Scribes and Pharisees	66	When to	121, 136
Sheds	77, 140, 144	Why	133
Suture	91	Warming Water	147
Texas Cattle	9, 12, 16, 134, 138	Yards	42
Tags	13, 16, 18	Young Bulls	136

www.ingramcontent.com/pod-product-compliance
Lightning Source LLC
Chambersburg PA
CBHW062215220526
45471CB00009B/3205